Marx Joyce
Hardy Austen
Defoe Melville Cooper Hugo
Abbott Chesterton Emerson Eliot
Montaigne Haggard Grimm
Stoker Machiavelli Molière
Carroll Christie Byron
Wilde Maupassant Schiller
Garnett Engels
Goethe Fitzgerald Hawthorne Kafka
Einstein Smith Hall
Cotton Dostoyevsky
Baum Kipling Doyle Willis
Dumas Henry Nietzsche
Leslie Flaubert Turgenev Balzac
Stockton Vatsyayana Crane
Burroughs Verne
Curtis Tocqueville Vinci
Homer Whitman Gogol Busch
Darwin Widger Tolstoy
Potter Freud Thoreau Twain Scott
Zola Plato Harte
Kant Jowett Lawrence Dickens Hesse
Stevenson Burton
Andersen Cervantes
London Descartes Voltaire
Poe Aristotle Wells Cooke
Hale James Hastings
Bunner Shakespeare Chambers Irving
Richter Ida
Doré Dante Shaw Benedict
Swift Chekhov Pushkin Alcott
Wodehouse
Newton

tredition®

tredition was established in 2006 by Sandra Latusseck and Soenke Schulz. Based in Hamburg, Germany, tredition offers publishing solutions to authors and publishing houses, combined with worldwide distribution of printed and digital book content. tredition is uniquely positioned to enable authors and publishing houses to create books on their own terms and without conventional manufacturing risks.

For more information please visit: www.tredition.com

TREDITION CLASSICS

This book is part of the TREDITION CLASSICS series. The creators of this series are united by passion for literature and driven by the intention of making all public domain books available in printed format again - worldwide. Most TREDITION CLASSICS titles have been out of print and off the bookstore shelves for decades. At tredition we believe that a great book never goes out of style and that its value is eternal. Several mostly non-profit literature projects provide content to tredition. To support their good work, tredition donates a portion of the proceeds from each sold copy. As a reader of a TREDITION CLASSICS book, you support our mission to save many of the amazing works of world literature from oblivion. See all available books at www.tredition.com.

 Project Gutenberg

The content for this book has been graciously provided by Project Gutenberg. Project Gutenberg is a non-profit organization founded by Michael Hart in 1971 at the University of Illinois. The mission of Project Gutenberg is simple: To encourage the creation and distribution of eBooks. Project Gutenberg is the first and largest collection of public domain eBooks.

Old Roads and New Roads

William Bodham Donne

Imprint

This book is part of TREDITION CLASSICS

Author: William Bodham Donne
Cover design: Buchgut, Berlin – Germany

Publisher: tredition GmbH, Hamburg - Germany
ISBN: 978-3-8472-1473-1

www.tredition.com
www.tredition.de

Copyright:
The content of this book is sourced from the public domain.

The intention of the TREDITION CLASSICS series is to make world literature in the public domain available in printed format. Literary enthusiasts and organizations, such as Project Gutenberg, worldwide have scanned and digitally edited the original texts. tredition has subsequently formatted and redesigned the content into a modern reading layout. Therefore, we cannot guarantee the exact reproduction of the original format of a particular historic edition. Please also note that no modifications have been made to the spelling, therefore it may differ from the orthography used today.

OLD ROADS
and
NEW ROADS.

"messer ludovico, dove avete cogliato tante coglionerie?"

LONDON:
CHAPMAN AND HALL, 193, PICCADILLY.

1852.

p. ii printed by
john edward taylor, little queen street,
lincoln's inn fields.

p. iiiPREFACE.

Gentle Reader,

If you look to move through this little volume in a direct line, after the present fashion of Railway Travelling, you will be signally disappointed. Nothing can well be more circuitous than the route proposed to you, nor more eccentric than your present guide. This book aspires to the precision of neither Patterson nor Bradshaw. Let men "bloody with spurring, fiery hot with speed," consult those oracles of swiftness and rectitude of way: we do not belong to their manor. We desire to beguile, by a sort of serpentine irregularity, the occasional tedium of rapid movement. We move to our journey's end by sundry old-fashioned circuitous routes. Grudge not, while you are whirled along a New Road, to loiter mentally upon certain Old Roads, and to consider as you linger along them the ways and means of transit which contented our ancestors. Although their coaches were slow, and their pack-saddles hard as those of the Yanguesan carriers of La Mancha, yet they p. ivreached their inns in time, and bequeathed to you and me—Gentle Reader—if we have the grace to use them, many pithy and profitable records of their wayfaring. The battle is not always to the strong, nor the race to the swift: neither is the most rapid always the pleasantest journey. Horace accompanied Mfcenas on very urgent business, yet he loitered on the way, and confesses his slackness without shame—

> "Hoc iter ignavi divisimus, altius ac nos
> Prfcinctis unum: minus est gravis Appia tardis."

It was, he says, more comfortable to take his time. Is our business more pressing than his was? It can hardly be, seeing that he wended with a company whose errand was to prevent the two masters of the world from coming to blows. In comparison with such a mission, who will put the buying of a cargo of cotton, or arriving an hour before a public meeting begins, or catching a pic-nic party just in the nick of time? St. Bernard rode from sunrise to sunset along the Lake Leman without once putting his mule out of a walk; so much delectation the holy man felt in beholding the beauty of the water and the mountains, and in "chewing the cud of his own sweet or bitter fancies." And good Michel Seigneur de Montaigne took a week for his journey from Nice to Pisa, although his horse was one of the smartest trotters in Gascony, merely for the pleasure he felt in following the by-lanes. And did not Richard Hooker receive from Bishop Jewell p. vhis blessing and his walking-staff, and yet with such poor means of speed he thought not of the weary miles between Exeter and Oxford, but trudged merrily with a thankful heart for the good oak prop, and the better blessing? Much less content with his journey was Richard when he rode to London on a hard-paced nag, that he might be in time to preach his first sermon at St. Paul's. And was not this, the hastier of his journeys, the most unlucky in his life, seeing that it brought him acquainted with that foul shrew, Joan, his wife, who made his after-days as bitter to him, patient and godly though he were, as wormwood and coloquintida? Are not these goodly examples, Christian and Heathen? Let the Train rush along, you and I will travel at our own pace.

Neither shall you, if you will be ruled by your present guide, saunter along the roads of Britain alone, or on known and extant ways only. Are there not roads which never paid toll, roads in the waste, roads travelled only in vision, roads once traversed by the feet of myriads, yet now overgrown by the forest, or buried deeply in the marsh? Shall we not for awhile be surveyors of these forgotten highways, and pause beside the tombs of the kings, or consuls, or Incas, who first levelled them? The world has moved westward with the daily motion of the earth. Yet, in the far East lie the most ancient highways—whose pavements once echoed with the hurrying feet of Nimrod's outposts p. vior the trampling of Agamemnon's rear-guard. It were well to mark how that ancient chivalry sped along their causeways.

Nor, on our devious route, shall baiting-places be wanting. Drunken Barnaby stayed not oftener to prove the ale than we will do:—

"Fgre jam relicto rure
Securem Aldermanni—bury
Primo petii, qua exosa
Sentina, Holburni rosa
Me excepit, ordine tali
Appuli Gryphem Veteris Bailey:

Ubi experrectum lecto
Tres Ciconias indies specto,
Quo victurus, donec fstas
Rure curas tollet mfstas:
Ego etiam et Sodales
Nunc *Galerum Cardinalis*
Visitantes, vi Minervf
Bibimus ad *Cornua Cervi.*"

Our inns may not always be found at the roadside; and we may possibly ever and anon seem to have missed the track altogether. Yet we will come into the main line in the end, and, I trust, part with kindly feelings, when the time has come for saying

                SISTE VIATOR.

p. 1 OLD ROADS AND NEW ROADS.

We have histories of all kinds in abundance,—and yet no good History of Roads. "Wines ancient and modern," "Porcelain," "Crochet work," "Prisons," "Dress," "Drugs," and "Canary birds," have all and each found a chronicler more or less able; and the most stately and imposing volume we remember ever to have turned over was a history of "Button-making:" you saw at once, by the measured complacency of the style, that the author regarded his buttons as so many imperial medals. But of roads, except Bergier's volumes on the Roman Ways, and a few learned yet rather repulsive treatises in Latin and German, we have absolutely no readable history. How has it come to pass that in works upon civilization, so many in number, so few in worth, there are no chapters devoted to the great arteries of commerce and communication? The p. 2 subject of roads does not appear even on that long list of books which the good Quintus Fixlein *intended* to write. Of Railways indeed, both British and foreign, there are a few interesting memorials; but Railways are one branch only of a subject which dates at least from the building of Damascus, earliest of recorded cities.

Perhaps the very antiquity of roads, and the wide arc of generations comprised in the subject, have deterred competent persons from attempting it; yet therefore is it only the more strange that incompetent persons have not essayed "this great argument," since they generally rush in, where their betters fear to tread. A history of roads is, in great measure indeed, a history of civilization itself. For highways and great cities not merely presuppose the existence of each other, but are also the issues and exponents of two leading impulses in the nature of man. Actuated by the one—the centripetal instinct—the shepherd races of Asia founded their great capitals on the banks of the Euphrates and the Ganges: impelled by the other— the centrifugal instinct—they passed forth from their cradle in the Armenian Highlands, westward as far as the Atlantic, and eastward as far as the Pacific. We have indeed indications of roads earlier than we have accounts of cities. For ages before Arcadian Evander came as a "squatter" to Mount Palatine, was there not the great road of the Hyperboreans from Ausonia to Delphi, by which, with p. 3 each revolving year, the most blameless of mankind conveyed to the Dorian Sun-god their offerings? And as soon as Theseus—the organizer of men, as his name imports—had slain the wolves and bears and the biped ruffians of the Corinthian Isthmus, did he not set up a direction-post, informing the wayfarer that "this side was Peleponnesus, and that side was Ionia"? Centuries of thought and toil indeed intervened between the path across the plain or down the mountain-gorge and the Regina Viarum, the Appian Road; and centuries between the rude stone-heap which marked out to the thirsting wayfarer the well in the desert, and the stately column which told the traveller, "This is the road to Byzantium."

In the land of "Geryon's sons," the paths which scaled the sierras were attributed to the toils of Hercules. In Bœotia, at a most remote era, there was a broad carriage-road from Thebes to Phocis, and at one of its intersections by a second highway the homicide of Laius opened the "long process" of woes, which for three generations enshrouded, as with "the gloom of earthquake and eclipse," the royal house of Labdacus. We have some doubts about the nature, or indeed the existence, of the road along which the ass Borak conveyed Mahommed to the seventh heaven: but we have no grounds for questioning the fact of the great causeway, which Milton saw in his vision, leading from Pandemonium to this earth, for have not Sin and Death p. 4been travelling upon it unceasingly for now six thousand years?

From that region beyond the moon, where, according to Ariosto—and Milton also vouches for the fact,—all things lost on earth are to be found, could we evoke a Carthaginian ledger, we would gladly purchase it at the cost of one or two Fathers of the Church. It would inform us of many things very pleasant and profitable to be known. Among others it would probably give some inkling of the stages and inns upon the great road which led from the eastern flank of Mount Atlas to Berenice, on the Red Sea. This road was in ill odour with the Egyptians, who, like all close boroughs, dreaded the approach of strangers and innovations. And the Carthaginian caravans came much too near the gold-mines of the Pharaohs to be at all pleasant to those potentates: it was

      —"much I wis
      To the annoyance of King Amasis."

But it is bootless to pine after knowledge irretrievably buried in oblivion. Otherwise we might fairly have wished to have stood beside King Nebuchadnezzar when he so unadvisedly uttered that proud vaunt which ended in his being condemned to a long course of vegetable diet. For doubtless he gazed upon at least four main roads which entered the walls of Babylon from four opposite quarters: —

> p. 5"From Arachosia, from Candaor east,
> And Margiana, to the Hyrcanian cliffs
> Of Caucasus, and dark Iberian dales:
> From Atropatia and the neighbouring plains
> Of Adiabene, Media, and the south
> Of Susiana, to Balsara's havens."

We pass over as a mad imperial whim Caligula's road from Baif to Puteoli, partly because it was a costly and useless waste of money and labour, and partly because that emperor had an awkward trick of flinging to the fishes all persons who did not admire his road. It was a bad imitation of a bad model—the road with which Xerxes bridled the "indignant Hellespont." Both the Hellespontine and the Baian road perished in the lifetime of their founders; while the Simplon still attests the more sublime and practical genius of Napoleon. We should have also greatly liked to watch the Cimbri and Ambrones at their work of piling up those gigantic earth-mounds in Britain and in Gaul, which, under the appellation of Devil's-dykes, are still visible and, as monuments of patient labour and toil, second only to the construction of the Pyramids.

The physiognomy of races is reflected in their public works. The warm climate of Egypt was not the only cause for the long paven corridors which ran underground from temple to temple, and conducted the Deputies of the Nomes to their sacerdotal meeting in the great Labyrinth. It was some advantage, indeed, to travel in the shade in a p. 6land where the summer heats were intense, and refreshing rains of rare occurrence; but it was a still greater recommendation to these covered ways that they enabled the priests to assemble without displaying upon the broad highway of the Nile the times and numbers of their synods. The pyramidal temples of Benares communicated by vaulted paths with the Ganges, as the chamber of Cheops communicated with the Nile. The capital of Assyria was similarly furnished with covered roads, which enabled the priests of Bel to communicate with one another, and with the royal palace, in a city three days' journey in length and three in breadth. Civilization and barbarism, indeed, in this respect met each another, and the caves of the Troglodyte Fthiopians on the western shore of the Red Sea were connected by numerous vaulted passages cut in the solid limestone, along which the droves of cattle passed securely in the rainy season to their winter stalls from the meadows of the Nile and the Astaboras.

Of the civil history of Carthage we know unfortunately but little. The colonists of Tyre and Sidon are to the ages a dumb nation. All we know of them is through the accounts of their bitter foes, the Greeks of Sicily and the Romans. It is much the same as if the only records of Manchester and Birmingham were to be transmitted to posterity by the speeches of Mr. George Frederic Young. Yet we know that the Carthaginians alone, among p. 7the nations of antiquity, made long voyages, — perchance even doubled the Cape three thousand years before Vasco de Gama broke the silence of the southern seas; and we are certain also that their caravan traffic with Central Africa and the coasts of the Red Sea passed along defined and permeable roads, with abiding land-marks of hostelry, well, and column. And we know more than this. The Romans, who jealously denied to other nations all the praise for arts or arms which they could withhold, yet accorded to the Carthaginians the invention of that solid intessellation of granite-blocks which is beheld still upon the fragments of the Appian Road. The highways which conveyed to the warehouses of Carthage the ivory, gold-dust, slaves, and aromatic gums of Central Libya ran through miles of well-ordered gardens and by hundreds of villas; and it was the ruthless destruction of these country-seats of the merchant-princes of Byrsa, which forced upon them the first and the second peace with Rome.

The Grecian roads, like the modern European highways, represented the free genius of the people: they were often sinuous in their course, and, respecting the boundaries of property, wound around the hills rather than disturb the ancient landmarks. Up to a certain point the character of the Grecian Republics was marked rather by rapid progression than by permanence. Their roads were of a less massive construction than the Roman, consisting for p. 8the most part of oblong blocks, and were not very artificially constructed, except in the neighbourhood of the great emporia of traffic, Corinth, and Athens, and Syracuse. Sparta possessed two principal military highways, one in the direction of Argolis, and another in that of Mycene; but the roads in the interior of Laconia were little better than drift-ways for the conveyance of agricultural produce from the field to the garner, or from the farm-yard to the markets of the capital and the sea-ports.

The Romans were emphatically the road-makers of the ancient world. An ingenious but somewhat fanciful writer of the present day has compared the literature of Rome to its great Vif. One idea, he remarks, possessed its poets, orators, and historians—the supremacy of the City on the Seven Hills; and Lucan, Virgil, Livy, and Tacitus, various as were their idiosyncrasies, still present a formal monotony, which is not found to the same degree in any other literature. This censure is, perhaps, as regards the literature of the Roman people, rather overstated; but it applies literally to their roads, aqueducts, and tunnels. The State was the be-all and the end-all of social life: the wishes, the prejudices, the conveniences of private persons never entered into account with the planners and finishers of the Appian Way, or the Aqueduct of Alcantara. The vineyard of Naboth would have been taken from him by a single *senat{s consultum*, p. 9without the scruples of Ahab and without the crime of Jezebel. The Roman roads were originally constructed, like our own, of gravel and beaten stone; the surface was slightly arched, and the Macadamite principle was well understood by the contractors for the earliest of the Sabine highways, the Via Salaria [9]. But after the Romans had borrowed from Carthage the art of intessellation, their roads were formed of polygonal blocks of immense thickness, having the interstices at the angles well filled with flints, and in some instances, as at Pompeii, with wedges of iron and granite; so that they resembled on a plane the vertical face of a Cyclopean or polygonal wall. Upon the roads themselves were imposed the stately and sonorous epithets of Consular and Prftorian; and had the records of the western Republic perished as completely as those of its commercial rival, the Appian Road would have handed down to the remotest ages one of the names of the pertinacious censor of the

Claudian house. To the Commonwealth, perpetually engaged in distant wars on its frontiers, it was of the utmost importance to possess the most rapid means of communicating with its provinces, and of conveying troops and ammunition. To the p. 10Empire it was no less essential to correspond easily with its vast circle of dependencies. The very life of the citizens, who, long before the age of Augustus, had ceased to be a corn-producing people, was sometimes dependent upon the facility of transit, and the rich plains of Lombardy and Gaul poured in their stores of wheat and millet, and of salted pork and beef, when the harvest of Egypt failed through an imperfect inundation of the Nile. But the convenience of travellers was as much consulted as the necessity of the subjects of Rome. A foot-pavement on each side was secured by a low wall against the intrusion or collision of wheel carriages. Stones to mount horses (for stirrups were unknown) [10] were placed at certain distances for the behoof of equestrians; and the miles were marked upon blocks of granite or peperino, the useful invention of the popular tribune Caius Gracchus. Trees and fences by the sides were cut to admit air, and ditches, like ours, carried off the rain and residuary water from the surface. The office of *Curator Viarum*, or Road Surveyor, was bestowed upon the most illustrious members of the Senate, and the Board of Health in our days may feel some satisfaction in knowing that Pliny the Younger once held the office of Commissioner of Sewers on the Fmilian p. 11Road. Nay, the ancients deemed no office tending to public health and utility beneath them; and after his victory at Mantinea, Epaminondas was appointed Chairman of the Board of Scavengers at Thebes.

We close this part of our subject, which must not expand into an archfological dissertation, with the following extract from the most eloquent and learned of the English historians who have treated of Rome.

> "All these cities were connected with one another and with the capital by the public highways, which, issuing from the Forum of Rome, traversed Italy, pervaded the provinces, and were terminated only by the frontiers of the empire. If we carefully trace the distance from the wall of Antoninus to Rome, and from thence to Jerusalem, it will be

found that the great chain of communication, from the north-west to the south-east point of the empire, was drawn out to the length of four thousand and eighty Roman miles. The public roads were accurately divided by milestones, and ran in a direct line from one city to another, with very little respect for the obstacles either of nature or of private property. Mountains were perforated, and bold arches thrown over the broadest and most rapid streams. The middle part of the road was raised into a terrace, which commanded the adjacent country, consisted of several strata of sand, gravel, and cement, and was paved with large stones, or, in some places p. 12near the capital, with granite. Such was the solid construction of the Roman highways, whose firmness has not entirely yielded to the effect of fifteen centuries. They united the subjects of the most distant provinces by an easy and familiar intercourse; but their primary object had been to facilitate the marches of the legions; nor was any country considered as completely subdued till it had been rendered in all its parts pervious to the arms and authority of the conqueror. The advantage of receiving the earliest intelligence, and of conveying their orders with celerity, induced the emperors to establish throughout their extensive dominions the regular institution of posts. Houses were everywhere erected at the distance only of five or six miles; each of these was constantly provided with forty horses, and by the help of these relays it was easy to travel a hundred miles on a day along the Roman roads."

Wherever the Romans conquered they inhabited, and introduced into all their provinces, from Syene, "where the shadow both way falls," to the *ultima Thule* of the Scottish border, the germs of Latin civilization. To this imperial people England and France owe their first roads; for the drift-ways along the dykes of the Celts scarcely deserve the name. The most careless observer must have remarked the strong resemblance between the right lines and colossal structure of the Roman Vif and the modern Railroad. We have indeed arrived at p. 13a very similar epoch of civilization to that of the Cfsarian era, but with adjuncts derived from a purer religion, and from more generous and expanded views of commerce and the interdependence of nations, than were vouchsafed by Providence to the ancient world.

Roads being so essential a feature of all political communities, it might have been expected that if no other feature of Roman cultivation had survived the wreck of the Empire, the great arteries of intercourse would at least have been retained. But the works of man's hand are the exponent of his ideas; and the ideas of the Teutonic and Celtic races who divided among themselves the patrimony of the Cfsars were essentially different from those entertained and embodied by Greece and Rome. The State ceased to be an organic and self-attracting body. The individual rather than the corporate existence of man became the prevalent conception of the Church and of legislators; and nations sought rather to isolate themselves from one another, than to coalesce and correspond. Moreover, the life of antiquity was eminently municipal. The city was the germ of each body politic, and the connection of roads with cities is obvious. But our Teutonic ancestors abhorred civic life. They generally shunned the towns, even when accident had placed them in the very centre of their shires or marks, and when the proximity of great rivers or p. 14the convenience of walls and markets seemed to hold out every inducement to take possession of the vacant enclosures. The castle and the cathedral became the nucleus of the Teutonic cities. Hamlets crept around the precincts of the sacred and the outworks of the secular building: but it was long before the Lord Abbot or the Lord Chatelain regarded with any feelings but disdain, the burgher who exercised his trade or exposed his wares in the narrow lanes of the town which abutted on his domains, and enriched his manorial exchequer.

In many cases indeed the Roman cities were allowed to decay: the forest resumed its rights: the feudal castle was constructed from the ruins of the Proconsul's palace and the Basilica, or if these edifices were too massive for demolition, they were left standing in the waste—the Mammoths and Saurians of a bygone civilization. The great Vif were for leagues overgrown with herbage, or concealed by wood and morass; and for the direct arms of transit which bound Rome and York together as by the cord of a bow, were substituted the devious and inconvenient highways, which led the traveller by circuitous routes from one province to another. The contrast indeed between the 'Old Road and the New' is represented in Schiller's fine image—rendered even finer in Coleridge's translation:—

> p. 15"Straight forward goes
> The lightning's path, and straight the fearful path
> Of the cannon ball. Direct it flies, and rapid,
> Shattering that it may reach, and shattering what it reaches.
> My son! the road the human being travels,
> That on which blessing comes and goes, doth follow
> The river's course, the valley's playful windings,
> Curves round the corn-field and the hill of vines,
> Honouring the holy bounds of property:
> And thus secure, though late, leads to its end."

It was long however before much security was found on the new roads. In the dark ages the days described by Deborah the prophetess had returned. "The highways were unoccupied, and the travellers walked through bye-ways: the villages were deserted. Then was war in the gates, and noise of the archers in the places of drawing water." Danger and delay were often the companions of the traveller. Occasionally a vigorous ruler, like Alfred, succeeded in restoring security to the wayfarer, and proved his success (so said the legend) by hanging up, in defiance of the plunderer, golden armlets on crosses by the roadside. But these intervals of safety were few and far between, and the traveller journeyed, like Coleridge's Ancient Mariner, "in fear and dread,"

> "Because he knew a fearful fiend
> Did close behind him tread."

The man-at-arms in the days of Border-war was a more formidable obstacle to progress than a wilderness of spectres. In the reign of Edward the Confessor the great highway of Watling Street p. 16was beset by violent men. If you travelled in the eastern counties, the chances were that you were snapped up by a retainer of Earl Godwin, and if in the district now traversed by the Great Northern Railway, Earl Morcar would in all likelihood arrest your journey, and without so much as asking leave clap a collar round your neck, with his initials and yours scratched rudely upon it, signifying to all men, by those presents, that in future your duty was to tend his swine or rive his blocks. Outlaws, dwelling in the forests or in the deep morass which girded the road, pounced upon the traveller on the causeway, eased him of his luggage if he carried any, and if there was no further occasion for his services, they either let him down easily into the next quagmire, or if they were, for those days, gentlemanly thieves, left him standing, as Justice Shallow has it, like a "forked radish," to enjoy the summer's heat or the winter's cold. The cross and escallop shell of the pilgrim were no protection: "Cucullus non fecit monachum" in the eyes of these minions of the road; or rather, perhaps, the hood gave a new zest to the wrongs done to its wearer by these "uncircumcised Philistines." Convents, the abodes of men professing at least to be peaceful, were obliged to keep in pay William of Deloraine to mate with Jock of Thirlstane: and ancient citizens were fain to put by their grave habiliments, and "wield old partisans in hands as old." There is extant an agreement made between Leofstan, p. 17Abbot of St. Albans, and certain barons, by which the Abbot agrees to hire, and the barons to let, certain men-at-arms for the security of the Abbey, and for scouring the forests. Savage capital punishments—impalement, mutilation, hanging alive in chains—were inflicted on the marauders, who duly acknowledged these attentions by yet more atrocious severities upon the wayfarers who had the ill luck to be caught by them.

The insecurity of the old roads necessarily affected the manners of the time. He should have been a hardy traveller who would venture himself "single and sole," when he might journey in company. The same cause which leads to the formation of the caravans of Africa and Asia, led to the collection of such goodly companies of pilgrims as wended their way from the Tabard in Southwark to the shrine of St. Thomas at Canterbury; and the pursuit of travelling under difficulties produced for all posterity the most delightful of the poems of the great father of English verse.

Travelling in companies, in times when it was next to impossible to be on "visiting terms with one's neighbours," tended greatly to the improvement of social intercourse, and to the erection of roomy and comfortable inns for the wayfarers. It took Dan Chaucer only a few hours to be on the best footing with the nine and twenty guests at the Tabard.

> p. 18"Befelle that, in that season [18] on a day,
> In Southwerk at the Tabard as I lay,
> Redy to wenden on my pilgrimage
> To Canterbury with devout corage,
> At night was come into that hostelrie
> Wel nine and twentie in a compagnie
> Of sondry folk, by aventure yfalle
> In felawship; and pilgrimes were they alle,
> That toward Canterbury wolden ride.
> The chambres and the stables weren wide,
> And wel we weren esed atth beste.
> And shortly, whan the sonne was gone to reste
> So hadde I spoken with hem everich on,
> That I was of hir felawship anon."

But the tenants of the waste and the woodland were not the only lords of the highway. The Norman baron drew little profit from the natural produce of his ample domains. In his way he was a staunch protectionist; but he left agriculture very p. 19much to take care of itself, and looked to his tolls, his bridges, and above all to his highways, for a more rapid return of the capital he had invested in accoutring men-at-arms, squires, and archers. We know, from 'Ivanhoe,' how it fared with Saxons, Pilgrims, and Jews, whose business led them near the castles of Front de Bœuf or Philip de Malvoisin: and we are certain that the Lady of Branksome kept, an expensive establishment, who were expected to bring grist to the mill of the lord or lady of the demesne, by turning out in all weathers and at all hours, whenever a herd of beeves or a company of pilgrims were descried by the watchers from Branksome Towers. For it must have taken no small quantity of beef and hides to furnish the Branksome retainers in dinners and shoe- and saddle-leather; since—

> "Nine and twenty knights of fame
> Hung their shields in Branksome Hall:
> Nine and twenty squires of name
> Brought them their steeds to bower from stall:
> Nine and twenty yeomen tall
> Waited duteous on them all:
> They were all knights of mettle true,
> Kinsmen to the bold Buccleugh."

When the traveller carried money in his purse, or the merchant had store of Sheffield whittles or Woodstock gloves in his pack, the lowest dungeon in the castle of the Bigods was his doom; and he was a lucky man who came out again from those crypts which now so much delight our archfological associations, with a tithe of his possessions, or with his proper allowance of eyes, hands, and ears.

Even on the Roman roads, with their good accommodation of pavement, milestones, and towns, journeys were for the most part performed on foot or horseback. For before steel springs were invented, it was by no means pleasant to ride all day in a jolting cart—and the most gorgeous of the Roman *carrucf*, or coaches, was no better. Pompous and splendid indeed—to pass for a moment from Norman and Saxon barbarism—must have been the aspect of the Queen of Roads within a few leagues of the capital of the world; splendid and pompous as it was to the actual beholder, it is perhaps seen to best advantage in the following description by Milton—

> "Thence to the gates cast round thine eye, and see
> What conflux issuing forth, or entering in;
> Prftors, proconsuls to their provinces
> Hasting or on return, in robes of state,
> Lictors and rods, the ensigns of their power,
> Legions and cohorts, turms of horse and wings;
> Or embassies from regions far remote,
> In various habits, on the Appian road,
> Or on the Fmilian."

As a pendant to this breathing picture oftan Old Road at the gate of the "vertex omnium civitatum," we subjoin a note from Gibbon:—

"The *carrucf* or coaches of the Romans were often of solid silver, curiously carved and engraved, and the trappings of the mules or horses were embossed with gold. This magnificence continued from the p. 21reign of Nero to that of Honorius: and the Appian Road was covered with the splendid equipages of the nobles who came out to meet St. Melania, when she returned to Rome, six years before the Gothic siege. Yet pomp is well exchanged for convenience; and a plain modern coach, that is hung upon springs, is much preferable to the silver and gold *carts* of antiquity, which rolled on the axle-tree, and were exposed, for the most part, to the inclemency of the weather." [21a]

The Anglo-Saxon generally travelled on horseback. The Jews were restricted to the ignobler mule. The former indeed had a species of carriage; and horse-litters, probably for the use of royal or noble ladies and invalids, are mentioned by Matthew Paris and William of Malmesbury. Wheel-carriages appear to have multiplied after the return of the Crusaders from Palestine—partly, it may be inferred, because increased wealth had inspired a taste for novel luxuries, and partly because the champions of the Cross had imbibed in the Holy War some of the prejudices of the infidels, and had grown chary of exposing to vulgar gaze their dames and daughters on horseback. [21b]

p. 22The speed of travelling depends upon the nature and facilities of the means of transit. Herodotus mentions a remarkable example of speed in a Hemerodromus, or running-post, named Phidippides, who in two days ran from Athens to Sparta, a distance of nearly 152 English miles, to hasten the Laconian contingent, when the Persians were landing on the beach of Marathon. Couriers of this order, trained to speed and endurance from their infancy, conveyed to Montezuma the tidings of the disembarkation of Cortes; and so imperfect were the means of communication at that era in Europe, that the Spaniards noted it as a proof of high refinement in the Aztecs to employ relays of running postmen, from all quarters of their empire to the city on the Great Lake. The speed of a Roman traveller was probably the greatest possible before the invention of carriage-springs and railways. We have some data on this head. The mighty Julius was a rapid traveller. He continually mentions his *summa diligentia* in his journal of the Gaulish Wars. The length of journeys which he accomplished within a given time, appears even to us at this day, and might well therefore appear to his contemporaries, truly astonishing. A distance of one hundred miles was no extraordinary day's journey for him. When he did not march with his army on foot,—as he often seems to have done, in order to set his soldiers an example, and also to express that sympathy with them which gained him their hearts so entirely—he mostly travelled in a *rheda*. This was a four-wheeled carriage, a sort of curricle, and adapted to the carriage of about half a ton of luggage. His personal baggage was probably considerable, for he was a man of most elegant habits, and sedulously attentive to his personal appearance. The tessellated flooring of his tent formed part of his *impedimenta*, and, like Napoleon, he expected to find amid the distractions of war many of the comforts and conveniences of his palace at Rome. He reached the Sierra Morena in twenty-three days from the date of his leaving Rome; and he went the whole way by land. The distance round the head of the Gulf of Genoa and through the passes of the Pyrenees is 850 leagues; and although the Carthaginians had once been masters of Spanish Navarre, the roads were far from regular or good. The same distance would now be accomplished in twelve days by a general and his mounted staff. From the usual rapidity with which the great Proconsul travelled, Cowley, in his Essay on 'Procrastination,' extracts a moral, or, as his Puritan contemporaries would have phrased it, a "pious use." "Cfsar," he says, "the man of expedition above all others, was so far from this folly (procrastination), that whensoever in a journey he was to cross any river, he never went out of his way for a bridge, or a ford, or a ferry, but flung himself into it immediately, and swam over; and this is the course we ought to imitate, if we meet with any stops in our way to happiness." In the time of Theodosius, Cfsarius, a magistrate of high rank, went post from Antioch to Constantinople. He began his journey at night, was in Cappadocia, 165 miles from Antioch, the ensuing evening, and arrived at Constantinople the sixth day about noon. The whole distance was 725 Roman, or 665 English miles.

Gibbon describes bishops as among the most rapid of ancient travellers. The decease of a patriarch of Alexandria or Antioch caused the death of scores of post-horses, from the rate at which anxious divines hurried to Constantinople to solicit from the Emperor the vacant see. On the whole however, in respect of speed in travelling, the Greeks and Romans were but slow coaches; and these exceptional instances merely serve to prove the general slackness of their pace. A Roman nobleman indeed, with all the means and appliances which his wealth could purchase, and with the positive advantage of the best roads in the world, travelled generally with such a ponderous train, that p. 25the heavy-armed legions with their parks of artillery might well advance as rapidly as an Olybrius or Anicius of the Empire. "In their journeys into the country," says Ammianus, "the whole body of the household marches with their master. In the same manner as the cavalry and infantry, the heavy and the light armed troops, the advanced guard and the rear, are marshalled by the skill of their military leaders; so the domestic officers who bear the rod, as an ensign of authority, distribute and arrange the numerous train of slaves and attendants. The baggage and wardrobe move in the front; and are immediately followed by a multitude of cooks and inferior ministers, employed in the service of the kitchen and of the table. The main body is composed of a promiscuous crowd of slaves, increased by the accidental concourse of idle or dependent plebeians."

At an even earlier period, in the age of Nero, before luxury had made the gigantic strides which distinguished and disgraced the Byzantine Court, Seneca records three circumstances relative to the journeys of the Roman nobility. They were preceded by a troop of Numidian light horse who announced by a cloud of dust the approach of a great man. Their baggage-mules transported not only the precious vases, but even the fragile vessels of crystal and *murra*, which last probably meant the porcelain of China and Japan. The delicate faces of the young slaves were covered with a medicated p. 26crust or ointment, which secured them against the effects of the sun and frost. Rightly did the Romans name their baggage *impedimenta*. A funeral pace was the utmost that could be expected from travellers so particular about their accommodations as these luxurious senators. Of a much humbler character was the state observed by the monarchs who succeeded to portions of the empire of the Cfsars. The Merovingian kings, when they employed wheel carriages at all, rode in wains drawn by bullocks; the Bretwaldas of the Saxon kingdoms went to temple or church on high festivals in the same cumbrous fashion; and "slow oxen" dragged the standard of the Italian Republics into the battle-field.

With the disuse or breaking up of the great Roman Vif in our island, the difficulty and delay of travelling increased, and more than thirteen centuries elapsed before it was again possible to journey with any tolerable speed. Wolsey indeed, it is well known, by the singular rapidity with which he conveyed royal letters to and from Brussels, galloped swiftly up the road of royal favour: and by his fast style of living at home afterwards galloped even more swiftly down again. Mordaunt, Earl of Peterborough, was noted for his incessant restlessness, and his rapid mode of passing from one land to another; but then he dispensed with all state and attendance, and rode like a post-boy from one end of Europe to another. As the readers p. 27of Pope, Swift, and their contemporaries are daily becoming fewer in number, we venture to extract the Dean's pleasant burlesque on this eccentric nobleman's migratory habits.

> "Mordanto fills the trump of fame,
> The Christian worlds his deeds proclaim,
> And prints are crowded with his name.
> In journeys he outrides the post,
> Sits up till midnight with his host,
> Talks politics and gives the toast;
> Knows every prince in Europe's face,
> Flies like a squib from place to place,
> And travels not, but runs a race.
> From Paris gazette `-la-main,
> This day arrived, without his train,
> Mordanto in a week from Spain.
> A messenger comes all a-reek,
> Mordanto at Madrid to seek;
> He left the town above a week.
> Next day the post-boy winds his horn,
> And rides through Dover in the morn;
> Mordanto's landed from Leghorn.

Mordanto gallops on alone;
The roads are with his followers strown;
This breaks a girth and that a bone.
His body active as his mind,
Returning sound in limb and wind,
Except some leather lost behind.
A skeleton in outward figure,
His meagre corpse, though full of vigour,
Would halt behind him, were it bigger.
So wonderful his expedition,
When you have not the least suspicion
He's with you like an apparition."

The badness of the roads and the rude forms of p. 28wheel-carriages added to the expense of travelling. A canon of Salisbury Cathedral may now travel to London at a cost which is scarcely felt by his prebendal income: but in the days of Peter of Blois the whole proceeds of a stall were inadequate to the expenses of such a journey. In the thirteenth century a bishop of Hereford was detained at Wantling by lack of money for post-horses, and but for the aid of some pious monastery or peccant baron in the neighbourhood, who seized the opportunity of compounding for his sins, the successor of the apostles must, like the apostles, have completed his journey on foot.

In the fourteenth century roads were so far improved, that jobbing horses became a regular business, and the licenses for hackneys and guides added to the returns of the exchequer. A fare of twelvepence was paid for horse-hire from Southwark to Rochester; and sixpence was the charge of conveyance from Canterbury to Dover. We do not know the rate at which the equestrians travelled. Ancient Pistol informs us that "the hollow-pampered jades of Asia could go but thirty miles a-day." But these cattle seem to have been like Jeshurun, fat and perchance kicking, and accustomed to the tardy pace of Asiatic pomp.

Shakspeare and Steele both expatiate on the casualties incident to riding upon hired horses. Petruchio and Catherine, like Dr. Samuel Johnson and Hetty, made their wedding tour on horseback; p. 29and each trip ended with a similar result—the temporary obedience of the fair brides to the marital yokes. After this fashion Grumio tells the story of the connubial ride:—"We came down a foul hill, my master riding behind my mistress." "Both on one horse?" says Curtis, apparently unacquainted with the fashion of pillions. "What's that to thee?" rejoins Grumio. "Tell thou the tale. But hadst thou not crossed me, thou shouldst have heard how her horse fell, and she under her horse; thou shouldst have heard in how miry a place; how she was bemoiled; how he left her with the horse upon her; how he beat me because her horse stumbled; how she waded through the dirt to pluck him off me; how he swore; how she prayed; how I cried; how the horses ran away; how her bridle was burst; how I lost my crupper."

That Petruchio rode a hired horse is rendered probable by the wretched character of his steed and its furniture. Hudibras or Don Quixote were not worse mounted than was the Shrew-tamer: seeing that his horse was "hipped with an old mothy saddle, the stirrups of no kindred; besides, possessed with the glanders, and like to mose in the chine; troubled with the lampass, infected with the fashions, full of wind-galls, sped with spavins, raied with the yellows, past cure of the fives, stark spoiled with the staggers, begnawn with the bots; swayed in the back and shoulder-shotten, near-legged before, and with a half-checked bit, p. 30and a headstall of sheep's leather; which, being restrained to keep him from stumbling, hath been often burst, and now repaired with knots; one girt six times pieced, and a woman's crupper of velure, here and there pieced with packthread." [30]

Steele (Tatler, No. 231) has borrowed, without any acknowledgement, from 'Taming the Shrew,' most of the circumstances of his story; yet his adoption of them shows that such a mode of travelling was still in common use in the seventeenth century. After the honey-moon was over, the bridegroom made preparations for conveying his new spouse to her future abode. But "instead of a coach and six horses, together with the gay equipage suitable to the occasion, he appeared without a servant, mounted on a skeleton of a horse which his huntsman had, the day before, brought in to feast his dogs on the arrival of their new mistress, with a pillion fixed behind, and a case of pistols before him, attended only by a favourite hound. Thus equipped, he, in a very obliging, but somewhat positive manner, desired his lady to seat herself on the cushion; which p. 31done, away they crawled. The road being obstructed by a gate, the dog was commanded to open it; the poor cur looked up and wagged his tail: but the master, to show the impatience of his temper, drew a pistol and shot him dead. He had no sooner done it, but he fell into a thousand apologies for his unhappy rashness, and begged as many pardons for his excesses before one for whom he had so profound a respect. Soon after their steed stumbled, but with some difficulty recovered; however, the bridegroom took occasion to swear, if he frightened his wife so again, he would run him through! And, alas! the poor animal, being now almost tired, made a second trip; immediately on which the careful husband alights, and with great ceremony first takes off his lady, then the accoutrements, draws his sword, and saves the huntsman the trouble of killing him: then says he to his wife, 'Child, prithee take up the saddle;' which she readily did, and tugged it home, where they found all things in the greatest order, suitable to their fortune and the present occasion." This veracious history proceeds to say that, after this practical lesson, the lady was ever remarkable for a sweet and compliant temper.

Cotton's—"cheerful hearty Mr. Cotton"—description of a post-horse may be less familiarly known to the reader than either of the preceding descriptions of the inconveniences of riding post: it describes a journey from the neighbourhood of Bakewell to Holyhead, about the year 1678.

> p. 32"A guide I had got, who demanded great vails,
> For conducting me over the mountains of Wales:
> Twenty good shillings, which sure very large is;
> Yet that would not serve, but I must bear his charges:
> And yet for all that, rode astride on a beast,
> The worst that e'er went on three legs, I protest.
> It certainly was the most ugly of jades:
> His hips and his rump made a right ace of spades;
> His sides were two ladders, well spur-galled withal;
> His neck was a helve, and his head was a mall;

For his colour, my pains and your trouble I'll spare,
For the creature was wholly denuded of hair,
And, except for two things, as bare as my nail, —
A tuft of a mane and a sprig of a tail.
Now such as the beast was, even such was the rider,
With head like a nutmeg, and legs like a spider;
A voice like a cricket, a look like a rat,
The brains of a goose, and the heart of a cat:
But now with our horses, what sound and what rotten,
Down to the shore, you must know, we were gotten;
And there we were told, it concerned us to ride,
Unless we did mean to encounter the tide.
And then my guide lab'ring with heels and with hands,
With two up and one down, hopped over the sands;
Till his horse, finding the labour for three legs too sore,
Foled out a new leg, and then he had four.
And now, by plain dint of hard spurring and whipping,
Dry-shod we came where folks sometimes take shipping.
And now hur in Wales is, Saint Taph be hur speed,
Gott splutter hur taste, some Welsh ale hur had need:
Yet surely the Welsh are not wise of their fuddle,
For this had the taste and complexion of puddle.
From thence then we marched, full as dry as we came,
My guide before prancing, his steed no more lame,
O'er hills and o'er valleys uncouth and uneven,
Until, 'twixt the hours of twelve and eleven,
More hungry and thirsty than tongue can well tell,
We happily came to St. Winifred's well."

Cotton's ride to Holyhead was not however p. 33nearly so diversified in its adventures as a journey from Hardwick to Bakewell about the same period, described by Edward, son of Sir Thomas Browne, the worthy knight and physician of Norwich.

A tour in Derbyshire, in the year 1622, was indeed no light matter. Our ancestors were much in the right to make their wills before encountering the perils of a ride across the moors. We are constrained to abridge the author's narrative, but the main incidents of it are preserved in our transcript.

> "This day broke very rudely upon us. I never travelled before in such a lamentable day both for weather and way, but we made shift to ride sixteen mile that morning, to Chesterfield in Derbyshire, passing by Bolsover Castle, belonging to the Earl of Newcastle, very finely seated upon a high hill; and missing our way once or twice, we rode up mountain, down dale, till we came to our inn, when we were glad to go to bed at noon. It was impossible to ride above two mile an hour in this stormy weather: but coming to our inn, by the ostler's help having lifted our crampt legs off our horses, we crawled upstairs to a fire, when in two hours' time we had so well dried ourselves without and liquored ourselves within, that we began to be so valiant as to think upon a second march; but inquiring after the business, we received great discouragement, with some stories of a moor, which they told us we must go over. We had by chance p. 34lighted on a house that was noted for good drink and a shovel-borde table, which had invited some Derbyshire blades that lived at Bakewell, but were then at Chesterfield about some business, to take a strengthening cup before they would encounter with their journey home that night. We, hearing of them, were desirous to ride in company with them, so as we might be conducted in this strange, mountainous, misty, moorish, rocky, wild country; but they, having

drank freely of their ale, which inclined them something to their countrie's natural rudeness, and the distaste they took at our swords and pistols with which we rid, made them loth to be troubled with our companies, till I, being more loth to lose this opportunity than the other (one of which had voted to lie in bed the rest of the day), went into the room and persuaded them so well, as they were willing, not only to afford us their company, but stayed for us till we accoutred ourselves. And so we most courageously set forward again, the weather being not one whit better, and the way far worse; for the great quantity of rain that fell, came down in floods from the tops of the hills, washing down mud, and so making a bog in every valley; the craggy ascents, the rocky unevenness of the roads, the high peaks, and the almost perpendicular descents, that we were to ride down: but what was worse than all this, the furious speed that our conductors, mounted upon such good horses, used to these hills, led us on with, p. 35put us into such an amazement, as we knew not what to do, for our pace we rode would neither give us opportunity to speak with them or to consult with one another, till at length a friendly bough that had sprouted out beyond his fellows over the road, gave our file leader such a brush of the jacket as it swept him off his horse, and the poor jade, not caring for its master's company, ran away without him: by this means, while some went to get his courser for him, others had time to come up to a general *rendezvous*; and concluded to ride more soberly: but I think that was very hard for some of these to do. Being all up again, our light-horsed companions thundered away, and our poor jades, I think, being afraid, as well as their masters, to be left alone in this desolate wide country, made so much haste as they could after them; and this pace we rid, till we lost sight of one another. At last our leaders were so civil, when it was almost too late, to make another halt at the top of one of the highest hills thereabout, just before we were to go to the moor: and I was the last that got up to them, where, missing one of my companions who was not able to keep up with us, I was in the greatest perplexity imaginable, and desiring them to stay awhile, I rid back again, whooping and hallooing out to my lost friend; but no creature could I see or hear of, till at last, being afraid I had run myself into the same inconvenience, I turned back again towards the mountaineers, p. 36whom when I had recovered, they told me 't was no staying there, and 't were better to kill our horses than to be left in those thick mists, the day now drawing to an end: and so setting spurs to their horses, they ran down a precipice, and in a short time we had the favour to be rained on again, for at the top of this hill we were drencht in the clouds themselves, which came not upon us drop by drop, but cloud after cloud came puffing over the hill as if they themselves had been out of breath with climbing it. Here all our tackling failed, and he that fared best was wet to the skin, these rains soaking through the thickest lined cloak: and now we were encountering with the wild moor, which, by the stories we had been told of it, we might have imagined a wild bore. I am sure it made us all grunt before we could get over it, it was such an uneven rocky track of road, full of great holes, and at that time swells with such rapid currents, as we had made most pitiful shift, if we had not been accommodated with a most excellent conductor; who yet, for all his haste, fell over his horse's head as he was plunging into some dirty hole, but by good luck smit his face into a soft place of mud, where I suppose he had a mouth full both of dirt and rotten stick, for he seemed to us to spit crow's nest a good while after. Now, being forced to abate something of their speed, I renewed my acquaintance with two of our new companions, and made them understand how we had left a third man behind us, not being able to ride so fast, and how our intentions were to stay at their own town with them this night, who now overjoyed to see an old acquaintance, were so kind and loving that what with shaking hands, riding abreast, in this bad way, and other expressions of their civilities, they put me in as much trouble with their favour as before they had put me to inconvenience by their rudeness: yet, by this means, I procured them to ride so easily as I led my horse down the next steep hill, on the side of which lay a vast number of huge stones, one intire stone of them being as big as an ordinary house: some of the smaller they cut into mill-stones. Passing the river—Derwent—which then ran with the strongest current that ever I beheld any, we climbed over another hill, a mile up and a mile down, and got to Bakewell a little after it was dark."

We have a few data of the speed possible in travelling on extraordinary occasions. We select one of each kind—that of the mounted express and that of the Great Lady who kept her carriage, as the extremes, so far as regards the instruments of conveyance. For a horseman can go where a wheel-carriage cannot find a track: and on the other hand, the traveller on foot can generally choose a more direct line of movement, than is practicable for the four-footed servant of man, encumbered with his rider and his furniture.

In the thirteenth and fourteenth centuries, the herald of the king of Scotland, who, it may be supposed, carried with him a royal mandate to be first served by the livery stables, was allowed forty days to reach the Border from London, although it appears that Robert Bruce took only seven to put the Border between himself and Gloucester. But neither Bruce nor the mother of Richard II., who came in one day from Canterbury to London, can be taken as precedents of ordinary speed. For the one had received a significant hint from some friendly courtier—a pair of spurs baked in a pie— that King Edward was in high dudgeon with him, and could not dine with either appetite or good digestion, until he had seen Bruce's head: and of the Queen dowager is it not written that "she never durst tarry on the waye," for Wat Tyler was behind her, vowing vengeance upon all principalities and powers? Howbeit her majesty was so thoroughly jolted and unsettled by the "slapping pace" at which she travelled, that she had a bilious attack forthwith, and was "sore syke, and like to die."

To the difficulty of transit on roads was owing the establishment of great annual fairs, still imperfectly represented by our Wakes, Statute-fairs, and periodical assemblages of itinerant vendors of goods. These commercial re-unions are still common in the East, and still frequent in Central Europe; although in England, where every hamlet has now happily its general shop, and where the towns p. 39rival the metropolis in the splendour of gas-lamps and the glory of plate-glass windows, such Fairs have degenerated into yearly displays of giants, dwarfs, double-bodied calves, and gorgeous works in gingerbread. To our ancestors, with their simpler habits of living, supply and demand, these annual meetings served as permanent divisions of the year. The good housewife who bought her woollens and her grocery, the yeoman who chose his frieze-coat, his gay waistcoat, and the leathern integuments of his sturdy props, once only in twelve months, would compute the events of his life after the following fashion:—"It happened three months after last Bury or Chester Fair;" or, "Please Heaven, the bullocks shall be slaughtered the week before the next Statute." Nay, dates were often extracted, in the courts of justice, by the help of such periodical memoranda. The Church of Rome, with its unerring skill in absorbing and insinuating itself into all the business or pleasures of mankind, did not overlook these popular gatherings. And if the ascetic Anthony, the sturdy Christopher, or that "painful martyr," St. Bartholomew, minded earthly matters in the regions of their several beatitudes, they must have been often more scandalized than edified by the boisterous amusements of those who celebrated their respective Feasts. In these particulars, however, Ecclesiastical Rome was merely a borrower from its elder Pagan sister. The Compitalia of ancient Rome were street-fairs p. 40dedicated to the worship of local deities, and the Thirty cities of Latium held annually, on the slope of the Alban Mount, a great fair as well as a great council of Duumviri and Decuriones. To the ancient fairs of Southern Italy we are indebted for one of our oldest and most agreeable acquaintances. The swinging puppets of the Oscans were gradually confined within a portable box, and danced or gesticulated upon a miniature stage. Their dumb-show was relieved by the extemporary jests and songs of the showman, until at length, one propitious morning, some Homer or Shakspeare of the streets conceived the sublime idea of embodying these scattered rays of satire and jest in the portly person of — Mr. Punch.

The original fair of the East and medifval Europe was one of the most instructive and picturesque spectacles among the many gatherings of the human race. The Great Fair of Novogorod assembled, and still continues to assemble, myriads of nearly every colour and costume: and in the market of "the Sledded Russ" the small-eyed Chinese stood side by side with the ebony-complexioned native of Guinea. Among the many pictures which Sir Thomas Browne desired to see painted was "a delineation of the Great Fair of Almachara in Arabia, which, to avoid the great heat of the sun, is kept in the night, and by the light of the moon." The worthy and learned knight does not mention the Great Fair of the *Hurdwar*, in the northern p. 41part of Hindostan, where a confluence of many millions of human beings is brought together under the mixed influences of devotion and commercial business, and, dispersing as rapidly as it has been evoked, the crowd "dislimns and leaves not a wrack behind." But fairs and general enterprise and opulence are not coeval: neither do they flourish in an age of iron roads and steam-carriages. In fact, they were the results of the inconvenience attendant upon travelling. It was once easier for goods to come to customers than for customers to leave their homes in search of goods. Inland trade was heavily crippled by the badness and insecurity of the highways. The carriages in which produce was conveyed were necessarily massive and heavy in their structure, to enable them to resist the roughness of the ways. Sometimes they were engulfed in bogs, sometimes upset in dykes, and generally they rolled heavily along tracks little less uneven than the roofs of houses.

As a direct result of these obstacles to speedy locomotion, the fruits of the earth, in the winter months, when the roads were broken up or flooded, were consumed by damp and worms in one place, while a few miles further on they might have been disposed of at high prices. Turf was burned in the stoves of London, long after coals were in daily use in the northern counties; and petitions were presented to the Houses of Parliament in the reign of Henry VIII., deprecating the destruction of growing p. 42timber for the supply of hearth-fuel. Nor were these miry and uneven ways by any means exempt from toll; on the contrary, the chivalry of the Cambrian Rebecca might have been laudably exercised in clearing the thoroughfares of these unconscionable barriers. It was a costly day's journey to ride through the domain of a lord abbot or an acred baron. The bridge, the ferry, the hostelry, the causeway across the marshes, had each its several perquisite. Exportation from abroad was oftener cheaper than production at home. It answered better to import cloth from Flanders than to weave and bring it from York: and land carriage from Norwich to London was nearly as burdensome as water-carriage from Lisbon. Coals, manure, grain, minerals, and leather were transported on the backs of cattle. An ambassador going or returning from abroad was followed by as numerous a retinue as if he had ridden forth conquering and to conquer. Nor were his followers merely for state or ceremony, but indispensable to his comfort, since the horses and mules which bore his suite carried also the furniture of his bed-room and kitchen, owing to the clumsiness of wheel-carriages. If, as was sometimes the case, a great lord carried half an estate on his back, he often consumed the other half in equipping and feeding his train: and among the pleasures utterly unknown to the world for more than five thousand years is, that both peer and peasant may now travel from Middlesex to any p. 43portion of the known world with only an umbrella and carpet-bag.

We have alluded in our sketch of the earliest roads to the general character of early travelling; but a few words in connection with roads remain to be said on that subject. Travelling for pleasure—taking what our grandfathers were wont to call the *Grand Tour*—were recreations almost unknown to the ancient world. If Plato went into Egypt, it was not to ascend the Nile, nor to study the monumental pictures of a land whose history was graven on rocks, but to hold close colloquy on metaphysics or divinity with the Dean and Chapter at Memphis. The Greeks indeed, fortunately for posterity, had an incredible itch for Egyptian yarns, and no sooner had King Psammetichus given them a general invitation to the Delta, than they flocked thither from Athens and Smyrna, and Cos and Sparta, and the parts of Italy about Thurium, with their heads full of very particular questions, and often, to judge by their reports of what they heard, with ears particularly open to any answers the Egyptian clergy might please to give. Yet pleasure was not the object of their journey. Science, as themselves said, curiosity, as their enemies alleged, was the motive for their encountering perils by land and water. Indeed we recollect only three travellers, either among the Greeks or Romans, who can properly be considered as journeying for pleasure. These were Herodotus—the prince of p. 44tourists, past, present, or to come,—Paullus Fmilius, and Cfsar Germanicus.

Herodotus, there is reason to suspect, did not himself penetrate far into Asia, but gathered many of his stories from the merchants and mariners who frequented the wine-shops of Ephesus and Smyrna. Considering the sources of his information, and the license of invention accorded to travellers in all ages, the Halicarnassian was reasonably sceptical: and generally warns his readers when he is going to tell them "a bouncer," by the words "so at least they told me," or "so the story goes." Paullus Fmilius travelled like a modern antiquary and connoisseur. And for beholding the master-pieces of Grecian art in their original splendour and in their proper local habitations, never had tourist better opportunities. A negotiation was pending between the Achfan League and the Roman Commonwealth; and since the preliminaries were rather dull, and Flaminius felt himself bored by the doubts and ceremonies of the delegates, he left them in the lurch to draw up their treaty, and took a holiday tour himself in the Peloponnesus. At that time not a single painting, statue, or bas-relief had been carried off to Italy. The Roman villas were decorated with the designs of Etrurian artists alone, or, at the most, had imported their sculpture and picture galleries from Thurii and Tarentum. Flaminius therefore gazed upon the entire mass of Hellenic art; and the only thing he, unfortunately p. 45for us, neglected, was to keep a journal, and provide for its being handed down to posterity.

Germanicus, who had beheld many of these marvels in the Forum and Palaces of Rome—for the Roman generals resembled the late Marshal Soult in the talent of appropriating what they admired—reserved his curiosity for Egypt alone, and traversed from Alexandria to Syene the entire valley of the Nile, listening complacently to all the legends which the priests deemed fitting to rehearse to Roman ears. He was of course treated with marked attention. Memnon's statue sounded its loudest chord at the first touch of the morning ray; the priests, in their ceremonial habiliments, read to him the inscriptions on the walls of the great Temple at Carnac—and proved to him that after all the Roman empire was no "great shakes;" since a thousand years before, Rameses III. had led more nations behind his chariot, and exacted heavier tributes of corn, wine, and oil from all who dwelt between the White Nile and the Caspian Sea. His journey however was so unprecedented a step, that it brought him into trouble with Tiberius. The Emperor was half afraid that Germanicus had some designs upon the kingdom of Egypt, and as that land happened to be the granary of Rome, the jealous autocrat thought of the possibility of short-commons and a bread-riot in the Forum. But even if the tourist had no ulterior views, the Emperor thought that it did not look like business for a proconsul to be making holiday without leave,—and he accordingly reprimanded his adopted son by letter, and scolded him in a speech to the senate. In our days the Emperor of Russia would look equally black on a field-marshal who should come without license to London for the season; and the Mandarin, who lately exhibited himself in the Chinese Junk, would do well for the future to eschew the Celestial Empire and its ports and harbours entirely,—at least if he have as much consideration for his personal comfort, as his sleek appearance indicated.

The Emperor Hadrian might have been added to the list of ancient travellers in search of the picturesque, both because he visited nearly every province of his empire, and because he expended good round sums wherever he went, in restoring, re-edifying, or beautifying the public edifices which the provincials had suffered to fall into decay. But Hadrian's journeys were primarily journeys of business; he wished, like the Czar Nicholas, to see with his own eyes how matters went on, and at times he had the felicity of catching a prefect in the very act of filling his pockets and squeezing the provincials: we cannot therefore put him to the account of those who journeyed for pleasure. Every Roman who took any part in public affairs was, in fact, a great traveller. If he served his sixteen or twenty years in the legions, and was not enrolled in the household troops, he was singularly unlucky p. 47if his company were not quartered in Asia, Africa, and the Danubian provinces. If he became prftor or consul, a provincial government awaited him at the close of his year of office; and it depended upon the billets drawn in the Senate, whether he spent a year or two on the shores of the Atlantic, or whether he kept staghounds on the frontiers of Dacia. Nearly every Roman indeed had qualified himself before he was fifty to be a candidate for the Travellers' Club; and sometimes the fine gentleman, who declined taking an active part in public affairs, found himself unexpectedly a thousand miles from home, with an imperial rescript in his portmanteau enjoining him not to return to Rome without special leave.

To such a compulsory journey was the poet Ovid condemned, apparently for his very particular attentions to the Princess Julia. His exile was a piece of ingenious cruelty. He was sent to Tomi, which was far beyond the range of all fashionable bathing-places. The climate was atrocious; the neighbourhood was worse; the wine was execrable and was often hard frozen, and eaten like a lozenge, and his only society was that of the barracks, or a few rich but unpolished corn-factors, who speculated in grain and deals on the shores of the Euxine. To write verses from morn to dewy eve was the unfortunate poet's only solace: and he sent so many reams of elegies to Rome, that his friends came at last to vote him a bore, p. 48and he was reduced, for want of fitting audience, to learn the Getic language, and read his lacrymose couplets to circles of gaping barbarians.

A few of our readers may remember the family coach in which county magnates rode in procession to church, to Quarter sessions, and on all occasions of ceremony and parade. The Landau, so fast disappearing from our streets and roads, was but a puny bantling of a vehicle in comparison with the older and more august conveyance. As the gentlemen rode on horseback, and the ladies upon pillions, on all but the great epochs of their lives, this wheeled mammoth was rarely drawn out of its cavern, the coach-house. For not even when in full dress, raised from the ground by red-heeled shoes resembling a Greek *cothurnus*, and with a cubit added to their stature by a mural battlement of hair, did the ladies of the eighteenth century disdain to jog soberly behind a booted butler with pistols in his holsters, and a Sir Cloudesley Shovel beaver on his head. [48] "We have heard an p. 49ancient matron tell of her riding nine miles to dinner behind a portly farm bailiff, and with her hair dressed like that of Madame de Maintenon, which, being interpreted, means that the locks with which nature had supplied her were further aggravated by being drawn tight over a leathern cushion—a fashion which Jonathan Oldbuck denounces as "fit only for Mahound or Termagaunt." The production of the coach was therefore the sign of a white or black day in the family calendar—inasmuch as it indicated either marriage or funeral, the approach of the Royal Judges or the execution of a state prisoner, the drawing for the militia, or a county address to both Houses of Parliament on the crying grievance of the Excise. It doubtless took some days to prepare the imperator's p. 50chariot for a Roman triumph: it must have employed nearly as many to clean and furbish the capacious body of the modern vehicle. There was moreover a whole armoury of harness to mend and polish; and as the six long-tailed Flemish horses were not often in the traces together, some time was required by them to unlearn the rustic habits of the farm-yard, and to regain the stately trot at which, where the roads would admit of it, they ordinarily proceeded. The following description of a journey to London by an M.P. of 1699 will convey to the reader a lively yet tolerably exact conception both of the glory and inconveniences of travelling in those days. It is taken from Vanbrugh's comedy of the 'Journey to London,' better known in its modern form of 'The Provoked Husband.'

"*James*. Sir, Sir, do you hear the news? They are all a-coming.

"*Uncle Richard*. Ay, Sirrah, I hear it.

"*James*. Sir, here's John Moody arrived already: he's stumping about the streets in his dirty boots, and asking every man he meets, if they can tell him where he may have a good lodging for a parliament-man, till he can hire such a house as becomes him. He tells them his lady and all the family are coming too; and that they are so nobly attended, they care not a fig for anybody. Sir, they have added two cart-horses to the four old geldings, because my lady will have it said she p. 51came to town in a coach and six—heavy George the ploughman rides postilion.

"*U. Richard.* Very well, the journey begins as it should do. Dost know whether they bring all the children with them?

"*James.* Only Squire Humphrey and Miss Betty, Sir; the other six are put to board at half-a-crown a week a head, with Joan Growse at Smoke-dunghill-farm.

"*U. Richard.* The Lord have mercy upon all good folks! What work will these people make! Dost know when they'll be here?

"*James.* John says, Sir, they'd have been here last night, but that the old wheezy-belly horse tired, and the two fore-wheels came crash down at once in Waggon-rut Lane. Sir, they were cruelly loaden, as I understand: my lady herself, he says, laid on four mail trunks, besides the great deal-box, which fat Tom sat upon behind.

"*U. Richard.* So!

"*James.* Then, within the coach there was Sir Francis, my lady, the great fat lap-dog, Squire Humphrey, Miss Betty, my lady's maid, Mrs. Handy, and Doll Tripe the cook; but she puked with sitting backwards, so they mounted her into the coach-box.

"*U. Richard.* Very well.

"*James.* Then, Sir, for fear of a famine before they should get to the baiting-place, there was such baskets of plum-cake, Dutch gingerbread, Cheshire p. 52cheese, Naples biscuits, maccaroons, neats' tongues and cold boiled beef; and in case of sickness, such bottles of usquebaugh, black-cherry brandy, cinnamon-water, sack, tent, and strong beer, as made the old coach crack again.

"*U. Richard.* Well said.

"*James.* And for defence of this good cheer and my lady's little pearl necklace, there was the family basket-hilt sword, the great Turkish scimitar, the old blunderbuss, a good bag of bullets, and a great horn of gunpowder.

"*U. Richard.* Admirable!

"*James.* Then for bandboxes, they were so bepiled up—to Sir Francis's nose, that he could only peep out at a chance hole with one eye, as if he were viewing the country through a perspective-glass."

The "blunderbuss, Turkish scimitar, and basket-hilt sword," in the foregoing extract from Vanbrugh, point to one of the constant perils of the road—the highwaymen. Lady Wronghead was lucky in bringing her "little pearl necklace" safe to London. Turpin's scouts, a few years later, would have obtained more accurate information of the rich moveables packed in the squire's coach. But as yet Turpin and Bradshaw were not. The great road from York to London however lay always under an evil reputation. It was by this line that Jeannie Deans walked to London, and verified the remark of her sagacious host, the Boniface p. 53of Beverley, that the road would be clear of thieves when Groby Pool was thatched with pancakes—and not till then. The example of Robin Hood was, for centuries after his death, zealously followed by the more adventurous spirits of Nottinghamshire, Leicestershire, and Yorkshire; and their enterprising genius was well seconded by the fine breed of horses for which those counties were famous. For cross-country work the Leicestershire blades had no fellows; and had the Darlington Hunt existed in those days, they would doubtless have been first a-field in the morning and last on the road at night. Nor were there any reasons in their dress, demeanour, or habits, why they should not consort with the best of the shire either when riding to cover, or celebrating the triumphs of the day afterward in the squire's hall, or the ale-house. Some of these redressers of the inequalities of fortune were of excellent houses,—younger sons, who having no profession—trade would have been disgraceful in their eyes—grew weary of an unvarying round of shooting, fishing, otter-hunting, and badger-baiting, and aspired, like their common ancestor Nimrod, to be hunters of men. Others had found the discipline of a regiment unpleasant, or had been unjust serving men. In short, the road, about a century and a half ago, was the general refuge of all who, like the recruits that flocked to King David at Adullam, were in distress or discontented. Mail-coach drivers and guards travelled armed to the teeth, booted to p. 54the hips, with bandeliers across their capacious chests, and three-cornered hats which, in conjunction with their flowing horse-hair wigs, were both sword- and bullet-proof. Passengers who had any value for their lives and limbs, when they booked themselves at London for Exeter or York, provided themselves with cutlasses and blunderbusses, and kept as sharp look-out from the coach-windows as travellers in our day are wont to do in the Mexican diligences. We remember to have seen a print of the year 1769 in which the driver of the Boston mail is represented in the armed guise of Sir Hudibras. He carries a horse-pistol in his belt, and a *couteau de chasse* slung over his shoulder, while the guard is accoutred with no less than three pistols and a basket-hilt sword, besides having a carbine strapped to his seat behind the coach. Between the coachman's feet is a small keg, which might indifferently contain "genuine Nantz" or gunpowder. One of the "insides," an ancient gentleman in a Ramilies wig, is seen through the capacious window of the coach affectionately hugging a carbine, and a yeoman on the roof is at once caressing a bull-dog, and supporting a bludgeon that might have served Dandie Dinmont himself. Yet all these precautions, offensive or defensive, were frequently of no avail: the gentlemen of the road were still better armed, or more adroit in handling their weapons. Hounslow Heath on the great western road, and Finchley Common p. 55on the great northern road, were to the wayfarers for many generations nearly as terrible as the Valley of the Shadow of Death. "The Cambridge scholars," says Mr. Macaulay, "trembled when they approached Epping Forest, even in broad daylight. Seamen who had just been paid off at Chatham were often compelled to deliver their purses at Gadshill, celebrated near a hundred years earlier by the greatest of poets as the scene of the depredations of Poins and Falstaff." The terrors of one generation become the sources of romance and amusement to later times. Four hundred years ago we should have regarded William of Deloraine as an extremely commonplace and inconvenient personage: he is now much more interesting than the armour in the Tower, or than a captain or colonel of the Guards. A century back we should have slept the more soundly for the knowledge that Jack Sheppard was securely swinging in chains; but in these piping times of peace his biography has extracted from the pockets of the public more shillings than the subject of it himself ever 'nabbed' on the king's highway. It is both interesting and instructive to observe how directly the material improvements of science act upon the moral condition of the world. As soon as amended roads admitted of more rapid movement from place to place, the vocation of the highway robber was at first rendered difficult, and in the end impossible to exercise on the greater thoroughfares. Fast p. 56horse-coaches were the first obstacle. Railways have became an insuperable impediment to "life on the road."

Charles Lamb indited one of his most pleasant essays upon the 'Decay of Beggars in the Metropolis.' In the rural districts vagrancy and mendicity still survive, in spite of constabulary forces and petty sessions. But the mendicity of the nineteenth century presents a very different spectacle from the mendicity of the seventeenth. The well-remembered beggar is no longer the guest of the parish-parson; the king's bedesmen have totally vanished; no one now supplicates for alms under a corporation-seal; nor is the mendicant regarded as second only to the packman as the general newsmonger of a neighbourhood. Who does not remember the description of foreign beggars in the 'Sentimental Journey'? Many of us have witnessed the loathsome appearance and humorous importunity of Irish mendicants. A century ago England rivalled both France and Ireland in the number of its professional beggars. In the days when travelling was mostly performed on horseback, the foot of the hills—the point where the rider drew bridle—was the station of the mendicant, and long practice enabled him to proportion his clamorous petitions to the length of the ascent. [56] The old soldier in 'Gil Blas' stood by p. 57the wayside with a carbine laid across two sticks, and solicited, or rather enforced, the alms of the passer-by, by an appeal to his fears no less than to his pity. The readers of the old drama will recall to mind the shifts and devices of the 'Jovial Beggars;'—how easily a wooden leg was slipped off and turned into a bludgeon; how inscrutable were the disguises, and how copious and expressive the slang, of the mendicant crew. Coleridge has justly described 'The Beggar's Bush' as one of the most pleasant of Fletcher's comedies; and if the Spanish novelists do not greatly belie the roads of their land, the mendicant levied his tolls on the highways as punctually as the king himself. Speed in travelling has been as prejudicial to these merry and unscrupulous gentry as acts against vagrancy or the policeman's staff. He should be a sturdy professor of his art who would pour forth his supplications on a railway platform; and Belisarius himself would hardly venture to stop a modern carriage for the chance of an *obolus*, to be flung from its window. A few of the craft indeed linger in bye-roads and infest our villages and streets; but *ichabod!*—its p. 58glory has departed; and the most humane or romantic of travellers may without scruple consign the modern collector of highway alms to the tender mercies of the next policeman and the reversion of the treadmill.

The modern highway is seldom in a direct line. A hill, a ford, or a wood sufficed to render it circuitous. All roads indeed through hilly countries were originally struck out by drivers of pack-horses, who, to avoid bogs, chose the upper ground. Roads were first made the subject of legislation in England in the sixteenth century: until then, they had been made at will and repaired at pleasure. A similar neglect of uniformity may be seen in Hungary and in Eastern Europe generally, even in the present day. The roads are made by each county, and as it depends in great measure upon the caprice or convenience of the particular proprietors or townships whether there shall be a road at all, or whether it shall be at all better than a drift-way or a bridle-track, it often happens that after bowling along for a score of miles upon a highway worthy of Macadam, the carriage of the traveller plunges into wet turf or heavy sand, merely because it has entered upon the boundary of a new county. Nay, even where the roads have been hitherto good, it often happens that the new Vicegespann, or Sheriff, a personage on whose character a good deal depends in county business, allows them to go to ruin for want of seasonable repairs. A similar p. 59irregularity was, in our own country, put a stop to in the reign of Mary, when it was enacted that each parish should maintain its own roads. A custom was borrowed from the feudal system: the lord of the manor was empowered to demand from his vassals certain portions of their labour, including the use of such rude implements as were then in use. The peasant was bound by the tenure of his holding, to aid in cutting, carting, and housing his lord's hay and corn, to repair his bridges, and to mend his roads. A portion of such services was, in the sixteenth century, transferred from the lord to the parish or the district; and the charges of repairing the highways and bridges fell upon the copyholder. He was compelled to give his labour for six days in the year, and his work was apportioned and examined by a surveyor. If this compulsory labour did not suffice, hired labour was defrayed by a parochial rate: and although the obligation is seldom enforced, yet it survives in letter in the majority of the court-rolls of our manors.

So entirely indeed was speed in travelling regarded by our ancestors as of secondary importance to safety and convenience, that even in journeying by a public coach the length of a day's journey was often determined by the vote of the passengers. The better or worse accommodation of the roadside inns was taken into account; and it was "mine host's" interest to furnish good ale and p. 60beef, since he was tolerably certain that, with such attractions within-doors, the populous and heavy-laden mail would not pass by the sign of the Angel or the Griffin. Long and ceremonious generally were the meals of our forefathers; nor did they abate one jot from their courtesies when travelling on "urgent business." On arriving at the morning or noontide baiting-place, and after mustering in the common room of the inn, the first thing to be done was to appoint a chairman, who mostly retained his post of honour during the journey. At the breakfast or dinner there was none of that indecorous hurry in eating and drinking which marks our degenerate days. Had the travellers affected such thin potations as tea and soup, there was ample time for them to cool. But they preferred the sirloin and the tankard; and that no feature of a generous reception might be wanting, the landlord would not fail to recommend his crowning cup of sack or claret. The coachman, who might now and then feel some anxiety to proceed, would yet merely admonish his fare that the day was wearing on; but his scruples would vanish before a grace-cup, and he would even connive at a proposal to take a pipe of tobacco, before the horn was permitted to summon the passengers to resume their places. Hence the great caution observable in the newspaper advertisements of coach-travelling. We have now before us an announcement of the kind, dated in the year 1751. It sets p. 61forth that, God willing, the new Expedition coach! will leave the Maid's Head, Norwich, on Wednesday or Thursday morning, at seven o'clock, and arrive at the Boar in Aldgate on the Friday or Saturday, "as shall seem good" to the majority of the passengers. It appears from the appellation of the vehicle, "the new Expedition," that such a rate of journeying was considered to be an advance in speed, and an innovation worthy of general notice and patronage. Fifty years before the same journey had occupied a week; and in 1664 Christopher Milton, the poet's brother, and afterwards one of King James II.'s justices, had taken eight-and-forty hours to go from the *Belle Sauvage* to Ipswich! At the same period the stage-coach which ran between London and Oxford required two days for a journey which is now performed in about two hours on the Great Western line. The stage to Exeter occupied four days. In 1703, when Prince George of Denmark visited the stately mansion of Petworth, with the view of meeting Charles III. of Spain, the last nine miles of the journey took six hours. Several of the carriages employed to convey his retinue were upset or otherwise injured; and an unlucky courier in attendance complains that during fourteen hours he never once alighted, except when the coach overturned or was stuck in the mud.

    Direction-posts in the seventeenth century were almost unknown. Thoresby of Leeds, the well-known p. 62antiquary, relates in his Diary, that he had well-nigh lost his way on the great north road, one of the best in the kingdom, and that he actually lost himself between Doncaster and York. Pepys, travelling with his wife in his own carriage, lost his way twice in one short hour, and on the second occasion narrowly escaped passing a comfortless night on Salisbury Plain. So late indeed as the year 1770 no material improvement had been effected in road-making. The highways of Lancashire, the county which gave to the world the earliest important railroad, were peculiarly infamous. Within the space of eighteen miles a traveller passed three carts broken down by ruts four feet deep, that even in summer floated with mud, and which were mended with large loose stones shot down at random by the surveyors. So dangerous were the Lancashire thoroughfares that one writer of the time charges all travellers to shun them as they would the devil, "for a thousand to one they break their necks or their limbs by overthrows or breaking down." In the winter season stage-coaches were laid up like so many ships during Arctic frosts, since it was impossible for any number of horses to drag them through the intervening impediments, or for any strength of wheel or perch to resist the rugged and precipitous inequalities of the roads. "For all practical purposes," as Mr. Macaulay remarks, "the inhabitants of London were further from Reading than p. 63they are now from Edinburgh, and further from Edinburgh than they are now from Vienna."

France generally is still far behind Britain in all the appurtenances of swift and easy travelling. In the eighteenth century it was relatively at par with this country. The following misadventures of Voltaire and two female companions, when on an excursion from Paris to the provinces, are thus sketched by the pen of Thomas Carlyle:—

> "Figure a lean and vivid-tempered philosopher starting from Paris, under cloud of night, during hard frost, in a large lumbering coach, or rather waggon, compared with which indeed the generality of modern waggons were a luxurious conveyance. With four starved and perhaps spavined hacks, he slowly sets forth under a mountain of bandboxes. At his side sits the wandering virago, Marquise du Chbtelet, in front of him a serving maid, with additional bandboxes, *et divers effets de sa mantresse.* At the next stage the postilions have to be beat up: they came out swearing. Cloaks and

fur-pelisses avail little against the January cold; 'time and hours' are the only hope. But lo! at the tenth mile, this Tyburn coach breaks down. One many-voiced discordant wail shrieks through the solitude, making night hideous—but in vain: the axle-tree has given way; the vehicle has overset, and marchionesses, chamber-maids, bandboxes, and philosophers are weltering in inextricable chaos. The carriage was in the stage next p. 64Nangis, about half-way to that town, when the hind axle-tree broke, and it tumbled on the road to M. de Voltaire's side. Madame du Chbtelet and her maid fell above him, with all their bundles and bandboxes, for these were not tied to the front but only piled up on both hands of the maid; and so, observing the law of gravitation and equilibrium of bodies, they rushed toward the corner where M. de Voltaire lay squeezed together. Under so many burdens, which half-suffocated him, he kept shouting bitterly; but it was impossible to change place; all had to remain as it was till the two lackeys, one of whom was hurt by the fall, could come up, with the postilions, to disencumber the vehicle; they first drew out all the luggage, next the women, and then M. de Voltaire. Nothing could be got out except by the top, that is, by the coach-door, which now opened upwards. One of the lackeys and a postilion, clambering aloft and fixing themselves on the body of the vehicle, drew them up as from a well, seizing the first limb that came to hand, whether arm or leg, and then passed them down to the two stationed below, who set them firmly on the ground."

It was not entirely for state or distinction of ranks that noblemen of yore were attended on their journeys by running footmen. A few supernumerary hands were needed in case of accidents on the road. A box of carpenters' tools formed an indispensable part of the baggage, and the p. 65accompanying lackeys were skilful in handling them, as well as in replacing the cast shoes of the horses, for many districts would not afford a Wayland Smith. The state of travelling was doubtless increased by these 'cursive appendages, bearing white wands, and decked in the gay liveries of the house which they served. In the 'Bride of Lammermoor' we have a graphic picture of these pedestrian accompaniments of the coaches of "Persons of Quality."

"The privilege of nobility in those days," says Sir Walter Scott, "had something in it impressive on the imagination: the dresses and liveries, and number of their attendants, their style of travelling, the imposing and almost warlike air of the armed men who surrounded them, placed them far above the laird who travelled with his brace of footmen; and as to rivalry from the mercantile part of the community, these would as soon have thought of imitating the state and equipage of the Sovereign. . . . Two running footmen, dressed in white, with black jockey caps, and long staves in their hands, headed the train; and such was their agility that they found no difficulty in keeping the necessary advance which the etiquette of their station required before the carriage and horsemen. Onward they came at an easy swinging trot, arguing unwearied speed in their long-breathed calling. Behind these glowing meteors, who footed it as if the avenger of blood had been behind them, p. 66came a cloud of dust, raised by riders who preceded, attended, or followed, the state-carriage."

In times when persons of quality journeyed in this stately and sumptuous fashion, it was often needful to mend the roads specially on their account. The approach of a Royal Progress, or the Lord Lieutenant of the county, was a signal for a general 'turn-out' of labourers and masons to lay gravel over the most suspicious places, and to render the bridges at least temporarily secure. Scarcely a Quarter sessions in the seventeenth century passed over without presentments from the Grand Jury against certain districts of the county; and few and favoured were the districts which escaped a good round fine from the Judges, as a set-off against the bruises and other damages which their Lordships sustained on their circuit. It was no unusual accident for the Court to be kept waiting many hours for the arrival of the Judge. Either his Lordship had been dug out of a bog, or his official wardrobe had been carried away by a bridgeless stream. Often, too, the patience of jurors was severely tried by the non-appearance of counsel. These inconveniences became more apparent after it had ceased to be the fashion for the Judges and the Bar to travel on horseback from one assize-town to another. Cowper, writing to his pedestrian friend Rose, playfully imagines that when he should attain to the dignity of the ermine, he would institute the practice of 'walking' p. 67the circuit. But equestrian circuits were long in use, and the Bar turned out as if their chase had been deer instead of John Doe and Richard Roe. When however it came to be thought indecorous for a Judge to wear jack-boots, the danger of wheel-carriages was sensibly felt by the luminaries of the law, and the periodical journeys of the votaries of Themis tended directly to the correction of ways as well as to the suppression of vice. A zealous High Sheriff or a loyal Lord Lieutenant would sometimes contribute out of his private purse to the security of the Bench: and the more enterprising towns began to think it concerned their honour that the delegates of Majesty should reach their gates scatheless and unwearied by the toils of the road.

But road-making entrusted to the separate discretion of parochial authorities was often performed in a slovenly, and always in an unsystematic, manner. In adopting a direct or a circuitous line of way innumerable predilections interfered, and parishes not rarely indulged in acrimonious controversies, especially when the time came for walking the boundaries. The dispute between broad and narrow gauges is indeed merely a modern form of a long-standing quarrel. A market-town and a seaport would naturally desire to have ample verge and room enough on their highways for the transport of grain, hides, and timber from the interior, and for carriage of cloth and manufactured p. 68or imported goods to the inland. On the other hand isolated parishes would contend that driftways were all-sufficient for their demands, and that they could house their crops or bring their flour from the mill through the same ruts which had served their forefathers. But in Charles II.'s reign, after the civil wars had given an impetus to the public mind, and while, although our foreign policy was disgraceful, and each cabinet more indecorous than its predecessor, the country at large was steadily advancing in prosperity, this lack of uniformity was acknowledged to be no longer tolerable. Compulsory labour and parochial rates, or hired labour and occasional outlays, were found alike insufficient to ensure good roads. An act was accordingly passed authorizing a small toll to pay the needful expenses. The turnpike-gate to which we are accustomed was originally a bar supported on two posts on the opposite sides of the road, and the collector sat, *sub dio*, at his seat of customs. It was long however before the advantages of this plan were acknowledged by the people. Riots, resembling the Rebecca riots, were of frequent occurrence in the less-frequented counties: the road-surveyor was as odious as the collector of the chimney-tax; the toll-bar was seen blazing at night; its guardian deemed himself fortunate to escape with a few kicks; and it was not until a much later day that a public or private coach could trundle along the roads without encountering deep p. 69and dislocating ruts, or rocking over a surface of unbroken stones. Frost and rain were more effective than the duly appointed surveyor in breaking up these rude materials, and reducing the surface to something resembling a level.

A few years since some of the most strenuous opponents of railways were to be found among the squirearchy. "Why," argued these rural magnates, "should our woods be levelled and our cornlands bisected, our game scared away and our parks disfigured by noise and smoke, to suit the convenience of the dingy denizens of Manchester, or the purse-proud merchants of Liverpool?" Similar arguments were urged not more than a century ago against the formation of new turnpike roads. The bittern, it was said, would be driven from his pool, the fox from his earth, the wild fowl would be frightened away from the marshes, and many a fine haunch of venison would be sent to London markets without the proper ceremonies of turning off and running down the buck. Merrie England could not exist without miry roads. In 1760 there was no turnpike road between the port of Lynn and the great corn and cattle market at Norwich. In 1762 an opulent gentleman, who had resided for a generation of mortal life in Lisbon, was desirous to revisit his paternal home among the meres of the eastern counties. His wish was further stimulated by the circumstance that his sister and sole surviving relative dwelt beside one of the p. 70great Broads, which, in these regions, penetrate far inland from the sea-coast. From London to the capital town of his native county his way was tolerably smooth and prosperous. The distance was about a hundred and ten miles, and by the aid of a mail coach he performed the journey in three days. But now commenced his real labours. Between his sister's dwelling and the provincial capital lay some twenty miles of alternate ridges of gravel and morass. Had he been a young man he might have walked safely and speedily under the guidance of some frugal swain or tripping dairymaid returning from market. Had he been a wise man he would have hired a nag, and trotted soberly along such bridle-roads as he found. But he was neither a young nor a wise man. His better years had been passed in the counting-houses of Santarem, and his bodily activity was impaired by long and copious infusions of generous old port. So, as he could neither walk nor ride, he deposited his portly and withal somewhat gouty person in a coach-and-six, and set forth upon his fraternal quest. He had little reason to plume himself upon the pomp and circumstance of his equipage. The six hired coach-horses, albeit of the strong Flanders breed, were in a few hours engulfed in a black pool; his coach, or rather his travelling mansion, was inextricably sunk in the same slimy hollow; and the merchant himself, whose journeys had hitherto been made on the sober back of a p. 71Lusitanian mule, was ignominiously dragged by two cowherds through his coach-windows,—and mounted on one of the wheelers, he was brought back, drenched and weary, to the place whence he set out. In high dudgeon, the purveyor of Bacchus returned to London, and could never be induced to resume the search of his "Anna soror."

Such imperfect means of transit materially affected both the manners and the intelligence of the age. Postal arrangements indeed existed, but of the rudest kind. It was common for letters to be left at the principal inns on the main road, to be delivered when called for. They remained often in the bar until the address was illegible, or smoke had dyed the paper a saffron-yellow. Special announcements of deaths and births or urgent business were necessarily entrusted to special messengers; and the title and superscription of these privately-sent letters generally contain very minute and even peremptory injunctions of a certain and swift delivery. But for such cautions, a rich uncle might have been quietly inurned without his expectant nephews hearing of his decease; and a whole college kept waiting, till the year of grace had passed, for the news of a fat rector's much-desired apoplexy. The death of good Queen Bess was not known in some of the remoter parishes of Devonshire until the courtiers of James had ceased to wear mourning for her. The Hebrews of York heard with quivering lips and ashen brows of the p. 72massacre of their people in London at Richard I.'s coronation, six weeks after it was perpetrated; and the churches of the Orkneys put up prayers for King James three months after the abdicated monarch had fled to St. Germain's. There was in nearly all rural districts the king of London and the king of the immediate neighbourhood. The Walpoles and Townshends in their own domains were far more formidable personages than George I.; and at a time when the King of Prussia's picture was commonly hung out at ale-house doors as an incitement to try the ale, [72] an ancient dame near Doncaster exclaimed, on being informed of his majesty's decease, "Lord a' mercy, is he! and, pray, who is to be the new Lord Mayor?"

A considerable improvement in the roads of Great Britain took place in the latter half of the preceding century. This change was partly owing p. 73to the advancing civilization of the larger towns and cities, and partly to the march of the Highlanders into England under Prince Charles Edward, in 1745. At that period communication was so imperfect that the Pretender had advanced a hundred miles from Edinburgh without exciting any peculiar alarm in the midland or southern counties, while in the metropolis itself no certain information could be obtained of the movements of the rebel army for some days after their departure southward. The Duke of Cumberland's march northward was much impeded by the difficulty of transporting his park of artillery. But after the decisive day of Culloden, the erection of Fort William, and the establishment of military posts at the foot of the Grampians, the expediency of readier communication between the capitals of South and North Britain was universally felt. Scotland could henceforward be held in permanent subordination only by means of good military highways. Accordingly in the year 1782 we find a German traveller (Moritz) speaking of the roads in the neighbourhood of London as "incomparable." He is astonished "how they got them so firm and solid;" and he thus describes his stage of sixteen miles from Dartford, the place of his disembarkation, to the metropolis: —

"Our little party now separated and got into two post-chaises, each of which held three persons, though it must be owned that three cannot sit quite so commodiously in these chaises as two; the hire of a post-chaise is a shilling for every English mile. They may be compared to our extra-posts, because they are to be had at all times. But these carriages are very neat and lightly built, so that you hardly perceive their motion, as they roll along these firm smooth roads; they have windows in front and on both sides; the horses are generally good, and the postilions particularly smart and active, and always ride at a full trot. The one we had wore his hair cut short, a round hat, and a brown jacket, of tolerably fine cloth, with a nosegay in his bosom. Now and then, when he drove very hard, he looked round, and with a smile seemed to solicit our approbation. A thousand charming spots and beautiful landscapes, on which my eye would long have dwelt with rapture, were now rapidly passed with the speed of an arrow."

It was one of Samuel Johnson's wishes that he might be driven rapidly in a post-chaise, with a pretty woman, capable of understanding his conversation, for his travelling companion. The smartness of the English postboy was emulated in France,—not, as might have been expected, by his professional brethren, who until very recently retained their ponderous jackboots, three-cornered hats, and heavy knotted whips, but by the younger members of *la haute noblesse*. To look like an English jockey or postilion, was long the object of fashionable ambition with Parisian dandies. "Vous me crottez, Monsieur," said poor patient Louis XVI. to one of these exquisite centaurs, as he rode beside the royal carriage near Versailles. "Oui, Sire, ` l'Anglaise," rejoined the self-satisfied dandy, understanding his majesty to have complimented his *trotting* (*trottez*), and taking it as a tribute to the skill of his imitation.

Pedlars and packhorses were a necessary accompaniment of bad and narrow roads. The latter have long disappeared from our highways; the former linger in less-frequented districts of the country, but miserably shorn of their former importance. A licensed hawker is now a very unromantic personage. His comings and goings attract no more attention among the rustics or at the squire's hall than the passing by of a plough or a sheep. The fixed shop has deprived him of his utility, and daily newspapers of his attractions. He is content to sell his waistcoat or handkerchief pieces; but he is no longer the oracle of the village inn or the housekeeper's room. In the days however when neither draper's nor haberdasher's wares could be purchased without taking a day's journey at the least through miry ways to some considerable market-town, the pedlar was the merchant and newsman of the neighbourhood. He was as loquacious as a barber. He was nearly as ubiquitous as the Wandering Jew. He had his winter circuit and his summer circuit. He was as regular in the delivery of news as the postman; nay, he often forestalled that government p. 76official in bringing down the latest intelligence of a landing on the French coast; of an execution at Tyburn; of a meteor in the sky; of a strike at Spitalfields; and of prices in the London markets. He was a favourite with the village crones, for he brought down with him the latest medicines for ague, rheumatism, and the evil. He wrote love-letters for village beauties. He instructed alehouse politicians in the last speech of Bolingbroke, Walpole, or Pitt. His tea, which often had paid no duty, emitted a savour and fragrance unknown to the dried sloe-leaves vended by ordinary grocers. He was the milliner of rural belles. He was the purveyor for village songsters, having ever in his pack the most modern and captivating lace and ribbons, and the newest song and madrigal. He was competent by his experience to advise in the adjustment of top-knots and farthingales, and to show rustic beaux the last cock of the hat and the most approved method of wielding a cane. He was an oral 'Belle Assemblie.' He was full of "quips and cranks and wreathed smiles." 'Indifferent' honest, he was not the less welcome for being a bit of a picaroon. Autolycus, the very type of his profession, — and such as the pedlar was in the days of Queen Bess, such also was he in the days of George II., — was littered under Mercury, and a snapper-up of unconsidered trifles. His songs would draw three souls out of one weaver. His pack was furnished with

p. 77 "Lawn, as white as driven snow;
Cyprus, black as e'er was crow;
Gloves, as sweet as damask roses,
Masks for faces and for noses;
Bugle-bracelet, necklace amber,
Perfume for a lady's chamber;
Golden quoifs and stomachers
For my lads to give their dears;
Pins and poking-sticks of steel, —
What maids lack from head to heel."

Then did he chant after the following fashion, at "holy-ales and festivals" —

"Will you buy any tape,
Or lace for your cape,
My dainty duck, my dear — a?
Any silk, any thread,
Any toys for your head,
Of the new'st and fin'st, fin'st wear — a?
Come to the pedlar,
Money's a meddler,
That doth alter all men's wear — a!"

One accident in pedlar life was some drawback to its general pleasantness. He often bore not only a great charge of goods, but of gold also. His steps were dogged by robbers, and many a skeleton, since disinterred in solitary places, is the mortal framework of some wandering merchant who had met with foul play on his circuit. The packman's ghost too is no unusual spectre in many of our shires.

How important a personage among the *dramatis personf* of rural life the pedlar was, at even a recent period, in the northern counties of England, may p. 78 be inferred from Wordsworth's choice of him for the hero of his 'Excursion.' Much ridicule, and even obloquy, did the staunch poet of Rydal incur for choosing such a character, when he might have taken Laras and Conrads by the score, and been praised for his choice. But "the vagrant merchant under a heavy load," being a portion of the mountain life which surrounded the poet's home, was better than any hero of romance for his purpose; and a younger generation has confirmed the poet's choice of a hero, and few remain now to mock at the Pedlar. Wordsworth's pedlar indeed was no Bryce Snailsfoot, nor Donald Bean, nor even such a one as was first cousin to Andrew Fairservice, but rather, by virtue of a poetic diploma, a philosopher of the ancient stamp. For

> "From his native hills
> He wandered far; much did he see of men,
> Their manners, their enjoyments and pursuits,
> Their passions and their feelings; chiefly those
> Essential and eternal in the heart,
> That, 'mid the simpler forms of rural life
> Exist more simple in their elements,
> And speak a plainer language. In the woods,
> A lone enthusiast, and among the fields,
> Itinerant in this labour, he had passed
> The better portion of his time; and there
> Spontaneously had his affections thriven
> Amid the bounties of the year, the peace
> And liberty of nature; there he kept
> In solitude and solitary thought,
> His mind in a just equipoise of love."

Lucian, in his vision of Hades, beheld the Shades p. 79of the Dead set by pitiless Minos or Rhadamanthus to perform tasks most alien to their occupations while they were yet denizens of earth. Nero, according to Rabelais, who improves on Lucian's hint, was an angler in the Lake of Darkness; Alexander the Great a cobbler of shoes; and "imperial Cfsar dead and turned to clay" a hawker of petty wares. It was easier to fit the shadows of monarchs with employment than it would be to find business for departed coachmen. "A coachman, Sir," said one of these worthies to ourselves, who was sorrowfully contemplating the approaching day of his extinction by a nearly completed railway,—"a coachman, if he really be one, is fit for nothing else. The hand which has from boyhood—or rather horsekeeper-hood—grasped the reins, cannot close upon the chisel or the shuttle. He cannot sink into a book-keeper, for his fingers could as soon handle a lancet as a pen. His bread is gone when his stable-door is shut." We attempted to console him by pointing out that it was a law of nature for certain races of mankind to become extinct. Were not the Red Men fading away before the sons of the

White Spirit? Was not the Cornish tongue, and were not the old Cornish manners, for ever lost to earth, on the day when the old shrewish fishwife, Dolly Pentrath, departed this life towards the middle of the reign of King George III.? Seeing these things are so, and that "all beneath the moon doth suffer change," why should p. 80coachmen endure for ever? But our consolation was poured into deaf ears, and some two years afterwards we recognized our desponding Jehu under the mournful disfigurements of the driver of a hearse. The days of pedlars and stage-coachmen have reached their eve, and look not for restoration. They are waning into the Hades of extinct races, with the sumpnours and the limitours of the Canterbury Pilgrims.

We have described some of the difficulties and dangers to which travellers were subjected in the days of Old Roads. Yet the ancient Highways were not without their attending compensations. Pleasant it was to travel in company, as Chaucer voucheth: pleasant to linger by the way, as Montaigne testifies. To meditative and imaginative persons there was delight in sauntering through a fair country, viewing leisurely its rivers, meadows, hills, and towns. Burton prescribes travelling among his cures for melancholy, and he would not have recommended railway speed or even a fast coach to sad and timid men. His advice presupposed sober progress, gliding down rivers, patient winding round lofty hills, contemplation by woodsides and in green meadows, relaxation not tension of nerve and brain. "No better physick," he says, "for a melancholy man than change of aire and variety of places, to travel abroad and see fashions. Leo Afer speakes of many of his countrymen so cured without all other physick. No man, saith p. 81Lipsius, in an epistle to Phil. Lanoius, a noble friend of his, now ready to make a voyage, can be such a stock or stone, whom that pleasant speculation of countries, cities, towns, rivers, will not affect. For peregrination charms our senses with such unspeakable and sweet variety, that some count him unhappy that never travelled, a kinde of prisoner, and pity his case, that from his cradle to old age beholds the same still; insomuch that Rhasis doth not only commend but enjoyn travell, and such variety of objects to a melancholy man, and to lye in diverse innes, to be drawn into severall companies. A good prospect alone will ease melancholy, as Gomesius contends. The citizens of

Barcino, saith he, are much delighted with that pleasant prospect their city hath into the sea, which, like that of old Athens, besides Fgina, Salamina, and other pleasant islands, had all the variety of delicious objects; so are those Neapolitanes and inhabitants of Genua, to see the ships, boats, and passengers go by, out of their windows, their whole cities being sited on the side of an hill like Pera by Constantinople. Yet these are too great a distance: some are especially affected with such objects as be near, to see passengers go by in some great road-way or boats in a river, *in subjectum forum despicere*, to oversee a fair, a market-place, or out of a pleasant window into some thoroughfare street to behold a continual concourse, a promiscuous rout, coming and going."

p. 82Indifferent roads and uneasy carriages, riding post, and dread of highwaymen, darkness or the inclemency of the seasons, led, as by a direct consequence, to the construction of excellent inns in our island. The superiority of our English hotels in the seventeenth century is thus described by the most picturesque of modern historians: — "From a very early period," says Mr. Macaulay, "the inns of England had been renowned. Our first great poet had described the excellent accommodation which they afforded to the pilgrims of the fourteenth century. Nine and twenty persons, with their horses, found room in the wide chambers and stables of the Tabard, in Southwark. The food was of the best, and the wines such as drew the company to drink largely. Two hundred years later, under the reign of Elizabeth, William Harrison gave a lively description of the plenty and comfort of the great hostelries. The continent of Europe, he said, could show nothing like them. There were some in which two or three hundred people, with their horses, could without difficulty be lodged and fed. The bedding, the tapestry, above all the abundance of clean and fine linen was matter of wonder. Valuable plate was often set on the tables. Nay, there were signs which had cost thirty or forty pounds. [82] In the seventeenth p. 83century, England abounded with excellent inns of every rank. The traveller sometimes in a small village lighted on a public-house, such as Walton has described, where the brick floor was swept clean, where the walls were stuck round with ballads, where the sheets smelt of lavender, and where a blazing fire, a cup of good ale, and a dish of trout fresh from the neighbouring brook, were to be procured at small charge. At the larger houses of entertainment were to be found beds hung with silk, choice cookery, and claret equal to the best which was drunk in London. The innkeepers too, it was said, were not like other innkeepers. On the continent the landlord was the tyrant of those who crossed his threshold. In England he was a servant. Never was an Englishman more at home than when he took his ease in his inn.

> "Many conveniences which were unknown at Hampton Court and Whitehall in the seventeenth p. 84century, are to be found in our modern hotels. Yet on the whole it is certain that the improvement of our houses of public entertainment has by no means kept pace with the improvement of our roads and conveyances. Nor is this strange; for it is evident that, all other circumstances being supposed equal, the inns will be best where the means of locomotion are worst. The quicker the rate of travelling, the less important is it that there should be numerous agreeable resting-places for the travellers. A hundred and sixty years ago a person who came up to the capital from a remote county generally required twelve or fifteen meals, and lodging for five or six nights by the way. If he were a great man, he expected the meals and lodging to be comfortable and even luxurious. At present we fly from York or Chester to London by the light of a single winter's day. At present therefore a traveller seldom interrupts his journey merely for the sake of rest and refreshment. The consequence is that hundreds of excellent inns have fallen into decay. In a short time no good houses of that description will be found, except at places where strangers are likely to be detained by business or pleasure."

Highwaymen, pedlars, inns, coachmen, and well-appointed coaches have now nearly vanished from our roads. Some of the more excellent breeds of English horses have gone with them, or will soon follow them. In another generation no p. 85one will survive who has seen a Norfolk hackney. This race of sure-footed indefatigable trotters has already become so few in number that "a child may count them." "The oldest inhabitant"—that universal referee with some persons on all disputed points—never set eye on a genuine Flemish coach-horse in England; and the gallant high-stepping hybrid—half thoroughbred, half hackney—which whirled along the fast coaches at the rate of twelve miles in the hour will in a few years be nowhere found. The art of 'putting to' four horses in a few seconds will become one of the 'artes deperditf;' and the science of driving so as to divide equally the weight and the speed between the team, and to apportion the strength of the cattle to the variations of the road, will have become a tradition. Perfect as mechanism was the discipline of a well-trained leader. He knew the road, and the duty expected of him. Docile and towardly during his seven- or nine-mile stage, he refused to perform more than his allotted task. Attached to his yoke-fellow, he resented the intrusion of a stranger into his harness: and a mere change of hands on the box would often convert the willing steed into a recusant against the collar, whom neither soothing nor severity would induce to budge a step. Some suffering indeed has been spared to the equine world by the substitution of brass and iron for blood and sinews; but the poetry of the road is gone with the *quadrigf* p. 86that a few years ago tripped lightly and proudly over the level of the Macadamized road. No latter-day Homer will again indite such a verse as

"Ἵππων μ' ὠκυπόδων ἀμφὶ κτύπος οὔατα βάλλει."

The Four-in-hand Club is extinct, or, with those ancient charioteers at Troy, courses in Hades over meadows of asphodel.

Of the old roads of the Continent during the Dark and Middle Ages, we have little to record. The central energy of Rome had suffered collapse. Europe was partitioned into feeble kingdoms and powerful fiefs. War was the normal condition of its provinces; the sports of the field were unfavourable to agriculture, and directly opposed to the promotion of commerce and the growth of towns. So long as it was conducive to the pleasures of the manorial lord to keep large tracts of land uncultivated, it was contrary to his interests to form great thoroughfares. We have in the 'Tesoretto' of Brunetto a striking picture of the desolation of northern Spain in the thirteenth century. He thus describes his journey over the plain of Roncesvalles.

> "There a scholar I espied,
> On a bay mule that did ride.
> Well away! what fearful ground
> In that savage part I found.
> If of art I aught could ken,
> Well behoved me use it then.
> More I look'd, the more I deem'd
> That it wild and desert seem'd:
> p. 87Not a road was there in sight;
> Not a house and not a wight;
> Not a bird and not a brute,
> Not a rill, and not a root;
> Not an emmet, not a fly,
> Not a thing I mote descry:
> Sore I doubted therewithal
> Whether death would me befall.
> Nor was wonder, for around
> Full three hundred miles of ground,
> Right across on every side
> Lay the desert bare and wide." [87]

As Ser Brunetto was despatched on very urgent business, it may be presumed that he was journeying by the most direct road which he could find. Until the reign of Charlemagne indeed there were but few towns, and consequently few roads, in Germany. The population generally was widely spread over the surface of the land. "A house, with its stables and farm-buildings," says Mr. Hallam, "surrounded by a hedge or inclosure, was called a court, or as we find it in our law-books, a curtilage: the toft or homestead of a more genuine English dialect. One of these, with the adjacent arable fields and woods, had the name of a villa or manse. Several manses composed a march; and several marches formed a Pagus, or district." There was indeed little temptation or need to move from place to place, when nearly p. 88every article of consumption was produced or wrought at home. For several centuries there is perhaps not a vestige to be discovered of any considerable manufacture. Each district furnished for itself its own articles of common utility. Rich men kept domestic artizans among their servants; even kings, in the ninth century, had their clothes made by the women upon their farms. The weaver, the smith, and the currier were often born and bred on the estate where they pursued their several crafts.

The position of Rome as the ecclesiastical metropolis of the world caused both a general and periodical recourse of embassies, deputations, pilgrims, and travellers to the Italian peninsula, yet we cannot discover that any especial conveniences were provided for the wayfarers. Even in the great and solemn years of the Jubilee the roads were merely patched up, and the bridges temporarily repaired by the Roman government, and only in such places as had become actually impassable. The floating capital of the more commercial of the Italian Republics was employed rather upon their docks and arsenals than upon their roads and causeways. Venice indeed, which for central vigour was the most genuine offspring of Imperial Rome, paved its continental possessions with broad thoroughfares. But neither Padua, Ravenna, nor Florence followed the example of the Adriatic Queen. On the contrary, Dante, when in his descent p. 89to Hell he meets with any peculiarly difficult or precipitous track, frequently compares it to some road well known to his countrymen, which fallen rocks had blocked up, or a wintry flood had rendered impermeable. Spain presented, as it presents at this day, to the engineer, almost insurmountable difficulties. The Moorish provinces of the south alone possessed any tolerable roads; nor were the ways of Arragon or Castile mended after the wealth of Mexico and Peru had been poured into the Spanish exchequer. Portugal owed its first good roads in modern times to its good king Emmanuel; and the Dutch and Flemings, the most commercial people of Europe from the thirteenth to the eighteenth centuries, found in their rivers and canals an easier transit than roads would have afforded them, for the wares which they brought from Archangel on the one hand and from the Spice Islands on the other. The military restlessness of France indeed led to the earlier formation of great roads. Yet France was a land long divided in itself; and the Duchies of Burgundy and Bretagne had little in common with the enterprising spirit of Paris, Lyons, and Marseilles. Upon the whole the roads of England, bad as they were, were at least upon a par with those of the Continent.

In this retrospect, hasty and imperfect as it is, we must not pass over the roads of Asia. And here ancient history affords us at least glimpses of p. 90definite knowledge. In that portion of the Asiatic continent which is seated between the Euxine Sea, the chain of Mount Taurus, and the Fgean, the crowded population, the activity of the Greek colonies, and the necessity for direct communication with the interior and seat of government, led to the construction of good and uniform highways. In the Ionian Revolt large bodies of troops were readily brought to bear upon the insurgents, and the preparations of Xerxes for his invasion of Greece cannot have been made without a previous provision of military roads. An exact scale of taxation was drawn up by Darius Hystaspes for all the provinces of his vast empire; and as the system survived the extinction of the royal house of Persia, and was adopted by the Macedonian conquerors in all its more important details, it may be inferred that such system worked with tolerable regularity and success. But as the tithes and tolls of Persia were paid both in money and in kind, it is obvious that the communication between the capital and satrapies of the empire must have been well organized. Such organization implies the existence of main roads radiating from Sousa and Ecbatana. Nor are we left to conjecture only. The establishment of running posts and couriers was a distinguishing feature of the Persian empire; and the speed at which they journeyed from the sea-coast or the banks of the Hyphasis to the seat of government proves that the roads were in good p. 91order and the stations and relays of runners well ascertained. The Anabasis of Cyrus—his "march up" the country—affords another proof. The narrative of Xenophon, in its earlier portions at least, and so long as the ten thousand Greeks kept to the main roads, resembles in the precision with which it marks distances and stations a Roman Itinerary or a Bradshaw's Guide. On this day, says the historical captain of mercenaries, we marched seven parasangs and bivouacked in an empty fort; on such a day we marched five parasangs and encamped in a pleasant park or 'paradise' of the great king. It is only after the Greeks have been forced from the 'Road-down' by the clouds of Persian cavalry, that they enter upon more rugged and devious mountain-paths. The account of Xenophon is confirmed by Arrian in his history of Alexander's Anabasis; and so long as the Macedonian conqueror was within the bounds of Persia proper, we rarely meet with any impediments to his progress arising out of the badness of the roads.

We have made some mention of the more conspicuous of ancient travellers. But travelling, either for business or pleasure, among the moderns, dates from the era of the Crusades. The barriers of the East were once again thrown open by that general ferment in the European world. Piety, the passion of enterprise, the dawning instincts of commerce, a new thirst for exotic luxuries, all contributed to inspire a desire for exploring the p. 92seats of the most ancient civilization. To this desire and to its effects we owe some of the most graphic and entertaining of modern writings. If we were, through any misadventure, sent to jail, we would stipulate for permission to carry into our cell Hakluyt's Voyages. The narratives of modern travellers are often learned, more often flimsy, and from the universality of locomotion, much given, like the prayers of the old Pharisees, to tedious repetitions. A tour in Greece or Italy now affects us with unutterable weariness. A journey from London to York affords more real novelty than many of these excursions. Sir Charles Fellows or Mr. Layard write in the spirit of the old travellers, and we would willingly wander any-whither with George Borrow. But, for the most part, the art of writing travels is lost—its imaginativeness, its credulity, its cherishing of mystery, and its proneness to awe. The old travellers are never sentimental—and sentiment is the very bane of road-books,—and they never describe for description's sake. The world was much too wonderful in their eyes for such unprofitable excursions of fancy. Beauty and danger, difficulty and strangeness, novel fashions and unknown garbs, were to them earnest and absorbing realities. The aspect of cities and havens, and leagues of forest and solitary plains, were to them "as a banner broad unfurled," and inscribed with mystic signs and legends. They were not whirled about from place to p. 93place: they had leisure to mark the forms and the colours of objects. They were in perils often: if they escaped shipwreck they were in danger of slavery; they journeyed with their lives in their hands, and were often yoke-fellows with hunger and nakedness, and the fury of the elements. Luckily for us who read their narratives, they were most unscientific, and ascribed the howling of the night-wind, the bursting of icebergs, the noise of tempests, and the echoes that traverse boundless plains after great heats, or are imprisoned in rock and fell, to the voice of demons exulting or lamenting to each other. We now cross the desert with nearly as much ease as we hail an omnibus, or book ourselves for Paris. But such was not the spirit in which Marco Polo, in the thirteenth, century, traversed the wilderness of Lop.

> "In the city of Lop," says the hardy and veracious merchant of Venice, "they who desire to pass over the desert cause all necessaries to be provided for them; and when victuals begin to fail in the desert, they kill their camels and asses and eat them. They mostly make it their choice to use camels, because they are sustained with little meat, and bear great burdens. They must purvey victuals for a month to cross it only, for to go through it lengthways would require a year's time. They go through the sands and barren mountains, and daily find water; yet at times it is so little that it will hardly suffice fifty or a hundred men with p. 94their beasts; and in three or four places the water is salt and bitter. The rest of the road, for eight-and-twenty days, is very good. In it there are not either beasts or birds; they say that there dwell many spirits in this wilderness, which cause great and marvellous illusions to travellers, and make them perish; for if any stay behind, and cannot see his company, he shall be called by his name, and so going out of the way be lost. [94a] In night they hear as it were the noise of a company, which, taking to be theirs, they perish likewise. Concerts of musical instruments are sometimes heard in the air, like noise of drums and armies. [94b] They go therefore close together, hang bells on their beasts' neck, and set marks, if any stray."

The Hebrews, dispersed over every region of the world, civilized or uncivilized, were necessarily p. 95great travellers. There was, in the first place, their central connection with Palestine, which they generally visited once in their lives, and whither thousands of them, as age advanced, flocked to lay their bones. There were the claims of kindred, prompting them to seek out and visit the children of dispersion, whether seated on the banks of the Vistula, the Euphrates, or the Nile; and there were the incentives of commerce, which drew them through the perils of land and sea. From the instructions given to their travelling agents in the medieval period, we derive much curious information respecting the internal state of Europe. It were indeed much to be wished that competent Hebrew scholars, instead of devoting themselves to the inane obscurity of the Rabbins, would employ their learning upon the history of the Jews in the Middle Ages. Much curious and interesting knowledge might be disinterred from the piles of Hebrew manuscripts that now lie amid the dust and spiders' webs of the Escurial. Above all things the itineraries of the Jewish travellers should be explored, as containing probably the most minute and accurate description of the social state of Europe at that period. Both for their personal security and for the despatch of their affairs, it was essential for the Jews to obtain and circulate the most exact information of the markets and population of the cities on their route. They required to know whom to shun and whom to seek; the towns in which the p. 96Jews' quarter was most commodious and secure; and the intervening tracts, often many days' journey in extent, which were most free from robbers or feudal oppressors. The following draft of instructions for a Spanish Jew, whose occasions led him from Spain to Greece, will afford the reader some conception of the historical value of such itineraries. Its date is apparently not later than the sixteenth century: —

> "Whoever wants to go from Saragossa, Huesca, Teruel, or any other town in Arragon, to Constantinople, the great city where the Turk reigns, must follow the route herein contained, and beware of the dangers that we are going to specify. The fugitive must first of all go to Jaca, where they will ask him the object of his voyage; he must say that he escapes to France, on account of his creditors, and he will not be disturbed. Thence he will go to Canfranc, and thence to Oleron, the first town in France, where, if questioned respecting the object of his voyage, he must say that he is going on a pilgrimage to Our Lady of Loretto. From Oleron to

Pau, to Tarbes, to Toulouse, to Gaillac, to Villefranche, and to Lyons: in this latter place the traveller will be obliged to show whatever money he carries, and pay one out of every forty pieces, whether silver or gold. At Lyons he will ask his way to Milan, and say that he is going to visit St. Mark of Venice; but when within five leagues of the former city, he must leave it on the right, and p. 97pass behind the mountain, so as not to enter the territory of the emperor. From thence he must direct his course towards the State of Venice; and when he arrives at Verona, not go through the city, for they make every one pay one real at the gates. In Verona he must ask his way to Padua, where he will embark on the river and go to Venice; the passage will cost him half a real. He will land on the Piazza di San Marco, and then he must look out for an inn to go to; he must be cautious in making his bargain with the innkeeper first; he must not pay more than half a real a day for his bed; and he is warned not to let the landlord provide him with anything, for he will charge him double for everything. On the day after his arrival he must go to the Piazza di San Marco, and there he will see some men with white turbans, and others with yellow; the first are Turks, the latter Jews. From these he will get every assistance and advice, whether he wants to go to Salonica or to any port of Greece."

At the time when Marco Polo, Rubruquis, Benjamin of Tudela, etc., journeyed in Asia, the East was still unspoiled—it was still the authentic Ophir of gold and barbaric pearl, and gorgeous armour, and solemn processions. At the same time Asia was but little behind Europe in the general elements of civilization, so that the contrast which is so glaring at the present day, between the state of a sultan and a pasha, and the squalid poverty of his p. 98subjects and servants, was then less startling. The courts of Europe were comparatively poor and mean, while the palaces of the oriental monarchs powerfully affected the imagination of the traveller. At a time too when the manners of the European nobility exhibited little refinement, the dignified courtesy and elaborate ceremonies of Bagdad and Ispahan were not less imposing than the pomp and splendour of their garb and its decorations. The Eastern chivalry also was to the full as efficient as that of the West; for what it lacked in weight of metal, it gained in superior adroitness in the use of weapons, in the greater facility of its movements, and the better temper and flexibility of its armour. All these features of a high—though, as it proved, a less enduring—civilization are noted with wonder and applause by the early travellers, who cannot sufficiently express their admiration of such opulence and such brilliant displays.

But for our immediate purpose, we can only speak of the great roads and inns of "Cathaian Khan." Marco Polo thus describes the great roads and excellent inns in the neighbourhood of Cambalu.

> "There are many public roads from the city of Cambalu, which conduct to the neighbouring provinces, and in every one of them, at the end of five-and-twenty or thirty miles, are lodgings or inns built, called *lambs*, that is, post-houses, with large and fair courts, chambers furnished with beds and other provisions, every way fit to entertain great p. 99men, nay, even to lodge a king. The provisions are laid in from the country adjacent: there are about four hundred horses, which are in readiness for messengers and ambassadors, who there leave their tired horses, and take fresh; and in mountainous places, where are no villages, the Great Khan sends people to inhabit, about ten thousand at a place, where these lambs or post-houses are built, and the people cultivating the ground for their provisions. These excellent regulations continue unto the utmost limits of the empire, so that, on the public ways throughout the whole of the Khan's dominions, about ten thousand of the king's inns are found; and the number of the horses appointed for the service of the messengers in those inns are more than two hundred thousand—a thing almost incredible: hence it is

that in a little while, with change of men and horses, intelligence comes, without stop, to the court. The horses are employed by turns, so that of the four hundred, two hundred are in the stables ready, the other two hundred at grass, each a month at a time. Their cities also, that are adjoining to rivers and lakes, are appointed to have ferry-boats in readiness for the posts, and cities on the borders of deserts are directed to have horses and provisions for the use of such as pass through those deserts: and they have a reasonable allowance for this service from the Khan. In cases of great moment the posts will ride two hundred miles a-day, or p. 100sometimes two hundred and fifty. Also they ride all night, foot-posts running by them with lights, if the moon does not shine.

"There are also between these inns other habitations, three or four miles distant from one another, in which there are a few houses, where foot-posts live, having each of them his girdle hung full of shrill-sounding bells. These keep themselves always ready, and as often as the Khan's letters are sent to them convey them speedily to the posts at the next village, who, hearing the sound of the foot-post coming when at a distance, expect him and receive his letters, and presently carry them to the next watch; and so, the letters passing through several hands, are conveyed, without delay, to the place whither they ought to come; and it often happens that by this the king learns news, or receives new fruits, from a place ten days' journey distant, in two days. As, for instance, fruits growing at Cambalu in the morning, by the next day at night are at Xanadu."

Such were the general features of the old roads of Asia and Europe centuries ago. But it must be regarded as one of the caprices of civilization that the only roads, in the fifteenth century, which rivalled the Roman Vif, were constructed in another hemisphere, and by a people whom the Europeans were wont to regard with disdain, as barbarous. The gold and silver furniture of the Peruvian palaces excited the cupidity of the Spanish p. 101invaders; but even avarice, for a moment, yielded to admiration, when the file-leaders of Pizarro's columns beheld for the first time the great Roads of the Incas. The Peruvians have been eloquently vindicated from the charge of barbarism by a modern historian, native of the great continent which Columbus discovered. From the moment when Cortes had gained the crest of the sierra of Ahualco, his progress was comparatively easy. Broad and even roads or long and solid causeways across the lakes and marshes conducted the Spaniards and their allies through the valley of Mexico or Tenochtitlan; and as they descended from the regions of sleet and snow, a gay and gorgeous panorama greeted them on every side, "of water, woodland, and cultivated plains," diversified with bold and shadowy hills, and studded with the roofs and towers of populous cities. The running posts of the Aztecs rivalled in speed and regularity their brethren in Cathay, and Montezuma could boast that his dominions displayed at least one element of civilization—rapid communication between the provinces and the capital—which in that age and long afterwards was unknown to the empire of his rival and conqueror, the 'white king beyond the seas.' The roads of Peru were however more wonderful than even those of Mexico. We now borrow Mr. Prescott's description.

> "Those," he says, "who may distrust the accounts of Peruvian industry, will find their doubts p. 102removed on a visit to the country. The traveller still meets, especially in the regions of the tableland, with memorials of the past, remains of temples, palaces, fortresses, terraced mountains, great military roads, aqueducts, and other public works, which, whatever degree of science they may display in their execution, astonish him by their number, the massive character of the materials, and the grandeur of the design. Among them, perhaps the most remarkable are the great roads, the broken remains of which are still in sufficient preservation to attest their former magnificence. There were

many of their roads traversing different parts of the kingdom; but the most considerable were the two which extended from Quito to Cuzco, and again diverging from the capital, continued in a southern direction towards Chili.

"One of these roads passed over the grand plateau, and the other along the lowlands on the borders of the ocean. The former was much the more difficult achievement, from the character of the country. It was conducted over pathless sierras buried in snow; galleries were cut for leagues through the living rock; rivers were crossed by means of bridges that swung suspended in the air; precipices were scaled by stair-ways hewn out of the native bed; ravines of hideous depth were filled up with solid masonry: in short, all the difficulties that beset a wild and mountainous region, and which might appal the most courageous engineer p. 103of modern times, were encountered and successfully overcome. The length of the road, of which scattered fragments only remain, is variously estimated at from fifteen hundred to two thousand miles, and stone pillars, in the manner of European milestones, were erected at stated intervals of somewhat more than a league, all along the route.

"The other great road of the Incas lay through the level country between the Andes and the ocean. It was constructed in a different manner, as demanded by the nature of the ground, which was for the most part low, and much of it sandy. The causeway was raised on a high embankment of earth, and defended on either side by a parapet or wall of clay; and trees and odoriferous shrubs were placed along the margin, regaling the sense of the traveller with their perfume, and refreshing him by their shade, so grateful under the burning sky of the tropics. In the midst of sandy wastes, which occasionally intervened, where the light and volatile soil was incapable of sustaining a road, huge piles were driven into the ground to indicate the route of the traveller."

Mr. Prescott might have added, that these magnificent works were constructed by a people ignorant of the use of iron, and unsupplied with wheel-carriages. The only beast of burden was the llama; and the long files of these patient and docile animals, winding along the broad causeways of the Andes recalled to the invaders the long p. 104strings of mules stepping in single file along the rocky paths cut out from the sides of the Iberian sierras. Iron and fire-arms alone were wanting to the Peruvians to enable them to rival the most potent of the European kingdoms both in the arts and arms which maintain empires.

Of New Roads we shall speak very briefly, and rather of their effects than of their history. It would indeed be idle, in a rapid sketch like the present, to be diffuse upon a subject which those who travel may study with their own eyes, and those who stay at home may learn from many excellent recent books. [104]

The defiance of natural obstacles, the massive piles of masonry, the filling up of valleys, the perforated hill, the arch bestriding the river or the morass, the attraction of towns towards the line of transit, the creation of new markets, the connection of inland cities with the coast, the interweaving of populations hitherto isolated from one another, the increase of land-carriage, the running to and fro of thousands whose fathers were born and died in the same town or the same district,—all these are features in common with the Flaminian and Fmilian ways, and with the roads laid down by the genius and enterprise of Stephenson. p. 105The old and the new roads, both in their resemblance and in their difference, suggest and express many of the organic distinctions and affinities of the old and the new phases of civilization.

For, apart from a feature of distinction already noticed, that in the ancient world all or nearly all public works were executed by and for the State, we may here remark that in England especially, where centralization is feeble, and local or personal interests are strong, the construction and conduct—the *curatio*, as the Romans phrased it—

of great roads are entirely in the hands of voluntary associations, and the State interferes so far only as to shield individual life and property from wanton wrong and aggression. Secondly, that the primary purpose of the Roman Vif was that of extending and securing conquest, while the primary end of the railroad is to diffuse and facilitate commerce. In the one case, civilization was a fortunate accident. Gaul imbibed the arts and manners of Latium, because Gaul had been first subdued, and was permanently held by the strong Roman arm. But, in the other case, traffic and communication are the direct objects, while war, if hereafter wars should arise, will be the crime or the infelicity of those who engage in it. War indeed, as all ancient history shows, was the normal condition of Heathendom; Peace, although so often in the past ages rudely interrupted, is the normal state of Christendom. Again, the Roman p. 106road rendered invasion, encroachment, and the lust of conquest easy to project, execute, and gratify; whereas the modern Vif, by bringing nations into speedy and immediate contact with one another, are diminishing with each year the chances of hostile collision. The Roman roads, with all their magnificent apparatus of bridges, causeways, of uplifted hollows and levelled heights, were constructed at an enormous cost of manual labour and of personal oppression and suffering, and with comparatively a trifling amount of science. But the railroad is the idea of the philosopher embodied by the free and cheerfully accorded toil of the labourer and artizan. When an Appius Claudius or a Marcus Flaminius determined to mark the year of his consulship or censorship by some colossal road-work, the husbandman was summoned from his field, the herdsman was brought from his pasture-ground, a contingent was demanded from the allies, a conscription was enforced upon the subjects of Rome, harder task-work was imposed on the slave, and more irksome punishment inflicted upon the prisoner. [107] The great works of antiquity indeed, from the pyramids downward to p. 107the mausoleum of Hadrian, are too often the monuments of human toil, privation, and death. But the roads of our more fortunate times are not cemented with the tears of myriads, nor reared upon piles of bleached bones. On the contrary, the construction of them has given employment to thousands who, but for them, would have crowded to the parish for relief, or have wandered anxiously in search of work, or sauntered listlessly at the alehouse door in despair of finding it. The great radii of peaceful communication have been executed by willing hands, and a fair day's wages has been the recompense of a fair day's work. We do not undervalue the skill and energy of the engineers of antiquity. Yet by their fruits we know and judge of the works of the Curatores Viarum, and of our Brunels and Stephensons. "Peace has its victories no less than war." And the modern road does not more surpass the ancient in the science of its constructors, than in the objects for which it has been planned and executed.

But before these results were attained, the air was tried, and the water was tried, as likely to afford a more rapid medium of transit and communication than the solid earth. Of balloons and canals however our limits do not permit us to speak, although either of them might well furnish a little volume like the one now presented to the reader. We are now concerned, however, with the social and civilizing effects of Railways.

"For a succession of ages," says Dr. Lardner, "the little intercourse that was maintained between the various parts of Great Britain was effected almost exclusively by rude footpaths, traversed by pedestrians, or at best by horses. Hills were surmounted, valleys crossed, and rivers forded by these rude agents of transport, in the same manner as the savage and settler of the backwoods of America or the slopes of the Rocky Mountains communicate with each other."

The effects of high civilization may perhaps be best estimated by its contrast—the rude and infant stages of society. Let us imagine for a moment the destruction of Railways, the neglect of Turnpike and Highway Roads, and the consequent interruption of our present modes of rapid and regular locomotion.

Gentle Reader, in the first place, your breakfast is rendered thoroughly uncomfortable, or, like Viola's history—a blank. Your copy of the 'Times' or 'Morning Chronicle' has not arrived; your letters are lying six miles off, and you have to send a special messenger—who may, and will most probably, get drunk on his road—to fetch them. If you should chance to be in business, you will hear of a profitable investment for capital just two hours after some one else has closed the bargain; if you are a physician, you will most probably miss a lucrative patient; if a lawyer, a most seductive fee. All calculations will be disturbed. Manchester and Norwich will be more remote from each other than Paris and Marseilles. In place of a railway station there will be a swamp, and instead of a turnpike gate, a wood. Mighty towns and spacious cities will shrink into obscure villages; smiling and fertile districts relapse into original barrenness; kinsfolk and acquaintance be put nearly out of sight. There are no mails; there is no penny post; the last new novel will not reach you. The Bishop of Exeter may become a cardinal, or Colonel Sibthorpe commander of the forces, six weeks before you hear of their promotion. The union between Scotland and England will be again as good as divorced by distance and difficulty of transit. Your fish from Billingsgate will be ancient, and your tailor will be sure to disappoint you of your mourning or your marriage suit. Your commodious carpet-bag must be exchanged for a trunk capacious enough to contain all your "household stuff," except the kitchen range; your utmost speed will amount to difficult stages of six miles an hour; you will journey in terror; and you will arrive at your inn with the fixed determination of never again quitting your home.

We will conclude our rambles over the old roads of four continents with the words of one whose wisdom was not surpassed by his wit, although his wit surpassed most of the wisdom of his contemporaries. p. 110"It is of some importance," says Sydney Smith, (it is wrong to add 'the Reverend,' for no one says *Mr.* William Shakspeare or *Mr.* John Milton,) "at what period a man is born. A young man alive at this period hardly knows to what improvement of human life he has been introduced; and I would bring before his notice the changes which have taken place in England since I began to breathe the breath of life—a period amounting to seventy years. Gas was unknown. I groped about the streets of London in all but utter darkness of a twinkling oil lamp, under the protection of watchmen in their grand climacteric, and exposed to every species of degradation and insult. I have been nine hours in sailing from Dover to Calais, before the invention of steam. It took me nine hours to go from Taunton to Bath before the invention of railroads, and I now go in six hours from Taunton to London! In going from Taunton to Bath I suffered between ten thousand and twelve thousand severe contusions, before stone-breaking Macadam was born. I paid fifteen pounds in a single year for repairs of carriage-springs on the pavement of London, and I now glide without noise or fracture on wooden pavement. I can walk, by the assistance of the police, from one end of London to the other without molestation; or, if tired, get into a cheap and active cab, instead of those cottages on wheels which the hackney coaches were at the beginning of my life. Whatever miseries I suffered, there p. 111was no post to whisk my complaints for a single penny to the remotest corners of the empire; and yet, in spite of all these privations, I lived on quietly, and am now ashamed that I was not more discontented, and utterly surprised that all these changes and inventions did not occur two centuries ago. I forgot to add that, as the basket of stagecoaches in which the luggage was then carried had no springs, your clothes were rubbed all to pieces; and that, even in the best society, one-third of the gentlemen at least were always drunk."

And now, Gentle Reader, have we not kept both troth and tryste with you? We put it to you seriously, did you ever chance to read a more rambling volume than the one now presented to you? You may talk to a pleasant companion in your first or second class carriage without losing the thread of our argument; you may indulge in a comfortable nap without its being necessary for you to mark the page where you dropped off. It may be better to begin at the beginning, and read in ordinary fashion to the close. But it will not be much worse if you have a fancy for commencing with the end. In short, you cannot go wrong, so you do but read in a charitable spirit—not being extreme to mark the much which is amiss.

Finally, we entreat of you to read this book in the temper which a certain English worthy recommends for his own.

> "One or two things yet I was desirous to have amended, if I could, concerning the manner of p. 112handling this my subject, for which I must apologize, *deprecari*, and upon better advice give the friendly reader notice. I neglect phrases, and labour wholly to inform my reader's understanding, not to please his ear. 'Tis my study to express myself readily and plainly as it happens: so that, as

a river runs, sometimes precipitate and swift, then dull and slow: now direct, then *per ambages*: now deep, then shallow: now muddy, then clear: now broad, then narrow; doth my style flow now serious, then light, as the present subject required, or as at the time I was affected. And if thou vouchsafe to read this Treatise, it shall seem to thee no otherwise than the way to an ordinary traveller, sometimes fair, sometimes foul; here champion, there enclosed; barren in one place, better soil in another. By woods, groves, hills, dales, plains, and lead thee *per ardua montium et lubrica vallium et roscida cespitum et glebosa camporum*, through variety of objects, to that which thou shalt like or haply dislike."

If thou art scholarly, Gentle Reader, running to and fro on Old or New Roads may do thee good. It will afford thee time to rest eye and hand, and furnish thee with more glimpses of this working world than are to be seen from a library-window. But if it chance that thou be not clerkly, then mayest thou both 'run to and fro' and 'increase thy knowledge' even with the aid of so poor a guide as he who now bids thee "Heartily Farewell."

Footnotes:

[9] The appellation of this, the earliest Roman road, affords another instructive example of the connection between the necessary wants of man and civilization. Salt, among the first needs of the city of Romulus, produced the path from the Salt-works; and the convenience of the Salt-work Road led ultimately to the construction of the Appian, Flaminian, and Fmilian.

[10] The first introduction of stirrups was probably not earlier than the end of the sixth century, a.d. See Beckmann's 'History of Inventions and Discoveries,' Eng. Trans., 1817, vol. ii. pp. 255-270.

[18] It is acknowledged on all hands that no people talk so much about weather as the English. It is also true that no literature contains so many descriptions of the sensations dependent on the seasons. A French or Italian poet generally goes to Arcadia to fetch images proper for "a fine day." We, on the contrary, paint from the life. Chaucer luxuriates, in his opening lines of the 'Canterbury Tales,' on the blessings and virtues of "April shoures." Our modern novelists are always very diffuse meteorologists. In lands where the seasons are unhappily uniform, the natives are debarred from this unfailing topic of conversation. Hajji Baba, in Mr. Morier's pleasant tale, is amazed at being told at Ispahan, by the surgeon of the English Embassy, that "it was a fine day." On the banks of the South American rivers, mosquitoes afford a useful substitute for meteorological remarks.—"How did you sleep last night?" "Sleep! not a wink. I was hitting at the mosquitoes all night, and am, you see, bitten like a roach notwithstanding."

[21a] The historian might have added to this description of Roman carriages an allusion to the sumptuousness of Roman harness. Apuleius informs us that "necklaces of gold and silver thread embroidered with pearls encircled the necks of the horses; that the head-bands glittered with gems; and the saddles, traces, and reins were cased in bright ribbons."

[21b] Not always, on horseback: for while the knight, as his Latin designation *eques* implied, was always mounted on a charger, his lady sometimes rode beside him on an ass:—

> "A loyely ladie rode him faire beside,
> Upon a lowly asse, more white than snow;
> Yet she much whiter; but the same did hide
> Under a vele, that wimpled was full low;
> And over all a black stole did she throw:
> As one that inly mourned so was she sad,
> And heavie sate upon her palfrey slow."

[30] We do not remember to have seen it remarked that Shakspeare has described all the good points of a horse, as well as (in the passage in the text) every imaginable bad one. The horse of Adonis was

> "Round-hoofed, short-jointed, fetlocks shag and long,
> Broad breast, full eye, small head, and nostril wide,
> High crest, short ears, straight legs, and passing strong,
> Thin mane, thick tail, broad buttock, tender hide."

[48] Riding as a Squire of Dames was occasionally a service of some danger. The long hair-pins which the ladies wore in their capillary towers were, as it appears from the following story, "as sharp as any swords." "Pardon me, good signor Don Quixote," says the duenna Donna Rodriguez to that unrivalled knight, "but as often as I call to mind my unhappy spouse, my eyes are brim-full. With what stateliness did he use to carry my lady behind him on a puissant mule, for in those days coaches and side-saddles were not in fashion, and the ladies rode behind their squires. On a certain day, at the entrance into St. James's Street in Madrid, which is very narrow, a judge of one of the courts happened to be coming out with two of his officers, and as soon as my good squire saw him—so well-bred and punctilious was my husband—he turned his mule about, as if he designed to wait upon him home. My lady, who was behind him, said to him in a low voice, 'What are you doing, blockhead? am I not here?' The Judge civilly stopped his horse and said, 'Keep on your way, Sir, for it is my business rather to wait on my lady Donna Casilda.' My husband persisted, cap in hand, in his intention to wait upon the Judge, which my lady perceiving, full of

choler and indignation, she pulled out a great pin and stuck it into his back; whereupon my husband bawled out, and, writhing his body, down he came with his lady to the ground. My mistress was forced to walk home on foot, and my husband went to a barber-surgeon's, telling him he was run quite through and through the bowels. But because of this, and also because he was a little short-sighted, my lady turned him away; the grief whereof, I believe, verily was the death of him."

[56] One of the most affecting of Wordsworth's pictures of rural manners is his sketch of the Old Cumberland Beggar. The opening lines of this excellent poem mark the usual station of the mendicant: —

> "I saw an aged Beggar in my walk;
> And he was seated by the highway side,
> On a low structure of rude masonry
> Built at the foot of a huge hill, that they
> Who lead their horses down the steep rough road
> May thence remount at ease."

[72] The practice of complimenting distinguished personages by suspending their portraits over ale-house doors sometimes indeed led to ludicrous consequences. We all remember the conversion of Sir Roger de Coverley's good-humoured visage into a frowning Saracen's Head. Soon after Dr. Watson had been installed at Llandaff, a rural Boniface exchanged for his original sign of the Cock an effigy of his new Diocesan. But somehow the ale was not so well relished by his customers as formerly. The head of the Bishop proved less inviting to the thirsty than the comb and spurs of the original Chanticleer. So to win back again the golden opinions of the public, mine host adopted an ingenious device. From reverence to the Church he retained the portrait of Dr. Watson, but as a concession to popular preferences he caused to be written under it the following inscription: —

> "This is the old Cock."

[82] The splendour and costliness of English signboards seem to have struck foreigners very forcibly. Moritz, from whom we have already quoted, says that "the amazing large signs which, at the entrance of villages, hang in the middle of the street, being fastened

to large beams, which are extended across the street from one house to another opposite to it, particularly struck me. These sign-posts have the appearance of gates, or gateways, for which I at first took them, but the whole apparatus, unnecessarily large as it seems to be, is intended for nothing more than to tell the inquisitive traveller that there is an inn." It marks in some degree the territorial prejudices of the English people that the principal inn of a hamlet usually "hangs out" the crest of the family, if it be indeed an ancient house, at the neighbouring hall or great house, whether it be a Swan, a Griffin, a St. George, or other heraldic or historic emblem or hero.

[87] We have availed ourselves of Mr. Cary's skilful translation of Brunetto's description of his journey from Florence to Valladolid, whither he had been sent on an embassy by the Guelph party: — "Un scolaio — Sur un muletto baio," etc.

[94a] It is perhaps scarcely necessary to observe how much indebted our great poets have been to the early travellers. Milton had perhaps this passage in his memory when he wrote the speech of the Lady in 'Comus': —

> "A thousand fantasies
> Begin to throng into my memory,
> Of calling shapes, and beck'ning shadows dire,
> And aery tongues, that syllable men's names
> On sands and shores and desert wildernesses."

[94b] "The isle is full of noises,
Sounds, and sweet airs, that give delight and hurt not.
Sometimes a thousand twangling instruments
Will hum about mine ears; and sometimes voices,
That, if I then had wak'd after long sleep,
Will make me sleep again." — *Tempest*, act iii. sc. 2.

[104] Among the most satisfactory of such works, we would especially mention 'A History of the English Railway,' by John Francis, in two volumes, 8vo, to which our own sketch is under great obligations.

[107] The staff of an ancient *Curator Viarum* resembled very nearly the accompaniments of a modern Railway contractor. "Caius Gracchus," says Plutarch, "was appointed supreme director for making

roads, etc. The people were charmed to see him followed by such numbers of architects, artificers, ambassadors, and magistrates: and he applied to the whole with as much activity, and despatched it with as much ease, as if there had been only one thing for him to attend to: insomuch that they who both hated and feared the man were struck with his amazing industry, and the celerity of his operations."

www.ingramcontent.com/pod-product-compliance
Lightning Source LLC
Chambersburg PA
CBHW030449220526
45464CB00006B/2464